T0198018

Science and Technology Research: Writing Strategies for Students

Tina M. Neville
Deborah B. Henry
Bruce D. Neville

The Scarecrow Press, Inc.
Lanham, Maryland, and Oxford
2002

SCARECROW PRESS, INC.

Published in the United States of America
by Scarecrow Press, Inc.
A Member of the Rowman & Littlefield Publishing Group
4720 Boston Way, Lanham, Maryland 20706
www.scarecrowpress.com

12 Hid's Copse Road
Cumnor Hill, Oxford OX2 9JJ, England

British Library Cataloguing in Publication Information Available

Library of Congress Cataloging-in-Publication Data Available

ISBN 0-8108-4429-X (cloth : alk. paper)
ISBN 0-8108-4428-1 (pbk. : alk. paper)

♾™ The paper used in this publication meets the minimum requirements of
American National Standard for Information Sciences—Permanence of
Paper for Printed Library Materials, ANSI/NISO Z39.48-1992.
Manufactured in the United States of America.

Contents

Figures

Preface

Information has now become a commodity. Students and researchers are presented with an overwhelming amount of information in a bewildering array of formats—printed journals and monographs, electronic journals and monographs, dissertations, technical reports, CD-ROMs, the World Wide Web, and more. Today's researchers need to become proficient in seeking out the precise information they need, and they must develop the ability to evaluate that information before it can be put to its intended use.

Science and Technology Research: Writing Strategies for Students is written to assist the reader in developing literature research and evaluation skills, without regard to format or source. It emphasizes basic principles of literature research and evaluation, avoiding as much as possible reliance on particular sources. Individual sources come, go, and change, so general principles that can be applied in any situation are considered more valuable than particular tips for specific sources of information. While the examples and skills are designed for an undergraduate seeking information in science and technology for the first time, the authors have endeavored to write the text so that it can be used by persons at any level of education, formal or otherwise, and in any field of study. The book can be used linearly by students for self-instruction or used as a resource for particular aspects of the library research process. It is also hoped that the book will prove useful to other library instructors as a textbook or as a resource for similar courses.

The authors of *Science and Technology Research: Writing Strategies for Students* are all experienced library instructors with backgrounds in science and/or engineering. While there are many books available on library skills, none specifically addresses researchers in science and technology, and very few focus on *general* research skills, focusing instead on specific information tools. In particular, few books concentrate on the important skills of refining a search question and evaluating the information retrieved. The authors recognized a need for a book that would provide an integrated development of the library research process from start to finish and concentrate on general skills that can be adapted to any library research problem.

The outline of the book is based on a popular for-credit library course taught at the University of South Florida, St. Petersburg, by Tina Neville and Deborah Henry. The book begins with an overview of information resources today, with emphasis on those available in most academic libraries, and the continuing importance of libraries and librarians. It then presents a general overview of the research process, including the essential process of establishing the question to be researched. General reference sources are covered, with emphasis on how they can assist in developing a research strategy, rather than concentrating on individual sources. Three chapters develop the skills needed to search electronic databases effectively. The importance of controlled vocabulary, Boolean commands, truncation and wildcards, and field searching to locate information precisely are discussed, again concentrating on principles that can be used *across* database platforms, rather than for specific platforms.

After the basic principles are mastered, advanced searching techniques are discussed for more difficult searching problems. The next six chapters cover a variety of general sources of information: on-line public access catalogs (OPACs) for books, journal and newspaper indexes, citation databases, bibliographies, government information, and the Internet. Following the discussion of the Internet, the all-important matter of *evaluation* of information is discussed to assist the reader in determining the relative reliability of the information gathered from the various sources. Four additional chapters discuss some tools for more specialized research: geographic information, statistical resources, ready-reference tools, and a variety of other sources. The book concludes with a sample research strategy, developed from beginning to end, that illustrates the principles developed throughout the book. It is hoped that this handbook will take the fear and mystery out of researching information and make it the productive and rewarding process that it can be.

Acknowledgments

It is our experience that today's college students are eager to learn about the various kinds of information that will help them in their research and in their daily lives. In formal and informal teaching sessions, we have been encouraged by their curiosity about the research and evaluation processes. We are indebted to the many students whose quest for knowledge has inspired us to write this book.

We would also like to thank Barbara Neville for taking the time to read the manuscript. She provided many helpful suggestions at a point when we were in danger of losing our focus.

Tina M. Neville
Deborah B. Henry
Bruce D. Neville

Chapter 1

Entering the Library World

Information is one of the most valuable commodities of the first part of the twenty-first century. Information is growing in volume, in complexity, in the variety of forms it takes, and in cost faster than anyone can keep up with it. At the same time, user's information needs are increasing in many of the same ways. Libraries are the traditional repositories of information, and they continue to fill that role. Libraries, as well as users, however, are struggling to keep up with the information explosion.

In the last decade of the twentieth century, libraries have experienced the most dramatic change since Gutenberg invented movable type. Just as Gutenberg's invention made books potentially accessible to anyone, computers and especially the Internet have vastly expanded the amount of information available to the average person. The row upon row of *card catalogs* in academic libraries a generation ago has been replaced by row upon row of computers. Most college students today have not personally experienced the days before photocopiers, on-line catalogs, e-mail, and electronic journals. The Internet has enabled libraries to provide research materials to their patrons anywhere, twenty-four hours a day, seven days a week. Some libraries are even discussing ways that they can provide users with twenty-four-hour access to a *reference librarian*.

In this book, we will deal primarily with academic libraries, those that serve colleges and universities. While some aspects of libraries and library users are unique to academic libraries, many principles remain the same across all types of libraries and users. We hope that this book will give you the skills to function in any type of library and with any type of information need.

The Academic Library

Academic libraries vary with the institutions they serve. Every library has its own feel, based on the people who work in it and the institution it serves. Smaller colleges and universities generally have a single library to serve all their users. As universities and their libraries grow, the library may be split into branches to serve specialized user populations. One of the first libraries to be split off is often the science and engineering library. As the size of the university grows still larger, even more specialized libraries, such as the engineering library or the life sciences library, may be split off. If the university has a medical school or a law school, there will generally be a separate library for those users. Scientists and engineers often need materials housed at the medical school library and may occasionally need to visit the law school library as well.

Like everything else about the academic world, academic libraries can be intimidating to the uninitiated. Almost certainly, your college or university library will be bigger than any you have experienced before, and the material will be much more specialized. Smaller college and university libraries may easily have a couple hundred thousand volumes; larger university library systems may have several million. The resources of the academic library are not limited to books. They also house journals, computer databases, microforms, government documents, audio and videotapes, materials that have been placed on reserve for a particular class—the list goes on and on. How are you supposed to find the particular information you need among the myriad resources suddenly at your disposal? Cheer up, there is hope, and this book is designed to provide you with the tools you will need to find your particular needle in a very large haystack.

Academic research relies heavily on primary material. In science and engineering, this generally means journals, but it also means technical reports, presentations at conferences and meetings and their subsequent proceedings, and a variety of other documents. In the science or engineering library, more space may be devoted to journals than to books. The number of scientific journals is staggering, and their cost even more so. A single scientific journal may have tens of thousands of pages a year and cost as much as a midsize automobile! Even with budgets in the millions of dollars a year, no library can afford to subscribe to every available journal. Each library therefore tailors its collection to the needs of its own patrons.

The Academic Librarian

The first rule for library users must be: *when in doubt, ask a librarian.* It is the reference librarian's job to connect users with the information they require. Most librarians entered the profession because of a sincere desire to help. Never apologize for interrupting the librarian at the reference desk. When things are slow at the desk, librarians may try to catch up on their own reading or paperwork, but the librarian's job while at the reference desk is to serve users; anything else is secondary.

Librarians are trained to know the resources available and to assist library users in finding the information they need. Reference librarians can answer many questions, but academic libraries also employ librarians who have developed an expertise in a particular field of study. In smaller college libraries, there may be only a single science librarian. In larger universities, there may be several—a chemistry librarian, an electrical engineering librarian, and so on. These librarians work with the faculty in the respective teaching and research departments to understand the information needs of the faculty and students in their subject areas, assist in selecting materials to add to the library collection, and provide instruction in the use of the library. Subject specialists may serve as the librarian of last resort for other reference librarians. They may also maintain Web pages of materials useful to patrons in their fields.

Library Skills

However academic libraries differ, the basics of information needs remain the same. Although the individual problems of users vary, many of the principles also remain the same. It is our goal in this book to provide you with the skills to identify your own information needs and to adapt them to the resources available.

One of the specialized functions of academic libraries and academic librarians is to teach their users the skills that they may need throughout their information-gathering lives—skills that you will learn in this book. While you are learning, reference librarians are there to guide you. They can help you develop skills in framing an appropriate research question, selecting an appropriate source to begin your search for information, developing a *search statement* that will retrieve relevant information, narrowing your search when you get too much information or expanding it when you get too little, and evaluating your information sources.

Computerization has had a major impact on library services and procedures, including reference, interlibrary loan, document delivery, and preservation. Computerization has greatly improved researchers' abilities to locate the particular information they need. Automation has also led to a certain level of intimidation and a great deal of frustration for library users. The same database may be provided by different services, each with a different set of commands. Rather than discuss particular systems, we have attempted to concentrate on general principles of retrieval. After completing this book, you should be able to determine when you need to perform more complex searching functions such as field or proximity searching. You should also know how to determine the commands you need to accomplish these functions in the particular version of the database that is available to you. Effective Internet searching is definitely a skill that requires practice and experience.

While computers have made many library tasks easier, faster, and more powerful, there are still skills that rely on more traditional technologies. Not all research materials are available in electronic formats, and not every institution can afford to provide those that are. Basic research in encyclopedias and ready-reference sources is often best conducted in print versions. Print sources are often most convenient for determining appropriate terms for computer searches. Most electronic databases do not extend back to the beginning of the indexes, so historical research may require access to the print volumes.

Perhaps the two most important research skills are the same in the computer and the traditional worlds: defining your research question and evaluating your resources. These two skills may have the greatest impact on the success of your research project, but are generally not given the attention they deserve. We will devote a chapter to each of these important processes. Again, a reference librarian can be one of your most valuable resources during these stages.

In appendix 1, we summarize ten important steps in library research for quick reference. Appendix 2 provides an actual example of a search strategy from beginning to end to demonstrate the techniques discussed throughout the book. Finally, always remember tip number 10 from appendix 1—have fun!

Chapter 2

Getting Focused

For many students, the most difficult part of a research paper or project may be trying to figure out exactly what to write about. Of course, if you are doing a literature search as background information for a laboratory experiment, it is quite easy to focus your research onto a particular chemical, organism, or process. Occasionally, your professor will assign to you a very specific subject, which may limit your flexibility. In most cases, though, you will probably have a good bit of leeway in your selections. If your professor wants you to write a ten-page paper relating to anything in your class, where do you begin to get ideas?

Initial Steps

Perhaps the most important first step, as you begin to think about potential topics, is finding a topic that is interesting to *you*. If you are not interested in the topic before you begin your research, you will definitely be bored by the time you finish!

The next thing to consider is whether or not the topic can be covered in the space that has been assigned. Your professor may specify a certain minimum or maximum length for your paper. It is a good idea to have a ballpark idea on how much depth you will need to include on your topic before you get too far into your research.

You should also consider whether the research can be concluded within the specified time limitations. If your paper is due in three days, you will probably need to use materials that are available at your local library or accessible on-line. If you have more time, you can think about *interlibrary loan* and *document delivery* options.

Finally, do you have the background knowledge that you will need to research the topic? We will be discussing ways to find definitions for unfamiliar terms or basic background on subjects that you may not have encountered; however, if the entire research project is based on principles that you have never learned, it may be worth postponing that topic until you have a better feel for the background concepts.

Selecting a Topic

A common trap that beginning researchers encounter is the tendency to select a topic that is too broad. Suppose you have been asked to write a paper for your botany class. Obviously, if you start looking up books and articles about "botany" you will find an overwhelming amount of information! You could narrow your search down a bit to "algae" or to "angiosperms," but the topic would still be much too general. Figure 2.1 illustrates how a very broad topic, botany, may be narrowed down into a much more specific, researchable topic, such as nitrogen fixation in the blue-green algae.

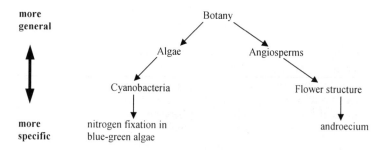

Figure 2.1. Diagram showing narrowing of a search topic.

There are many ways that you can find subjects for a research paper. Your course textbook may provide some ideas. Subject-specialized encyclopedias are also good sources for topics. General handbooks or bibliographies on a broader topic may also help you to generate ideas. Once you have a general idea, you can use a subject-specialized *the-*

saurus or the *Library of Congress Subject Headings (LCSH)* to help you find narrower and related terms. This will help you to focus your search. We will discuss how to use subject thesauri and the *LCSH* in chapter 4.

If your topic appears to be too broad, you may consider narrowing it down in several ways. For example, if you are looking for information on a particular process, such as digestive or reproductive systems, limit the research to a particular population, one species, or a small group of organisms. Instead of looking at the digestive system of all mammals, try studying the digestive system of cetaceans.

Another way to limit a search is to consider the time period that you need to cover. If researching a new technology, you could limit your search to information that has been published in the last two to five years. On the other hand, if you were studying the taxonomy of an organism, you may need to concentrate on historical information back to the earliest descriptions of a species.

Limiting the search to a particular geographic area is a third way to narrow a topic. A paper on cactus would be quite general, but an article on cactus species in Southern California might be much more workable to discuss in the space allocated to you.

Checking Your Resources

Once you have decided on a topic that you think meets all of your criteria for length, timeliness, and interest, you will want to be sure that there is enough information available on your topic to be able to conduct a literature search. You may have thought up the best topic in the world, but if no one else has studied it you won't find any published information on it. By the way, if you do come up with a topic like that, be sure to keep it in mind since it might make a great research project for graduate work or later publication!

For the purposes of this book, however, we are looking for information that we can use as background for a lab project or for a literature review on a particular topic. In those cases, you will want to be able to locate published information to support your theories. Before getting too far along into your research, check with your professor and with your academic reference librarian to see if they think that there will be adequate information available on your selected topic. It is useful to check the *catalog* at your library and an appropriate periodical index to see if you can locate information on your topic. Chapters 7 and 8 provide information on how to search these resources.

Focusing Your Search Statement

It is a good idea to re-formulate your general topic into a single question, hypothesis, or sentence. This helps to focus the topic in your own mind, and will improve your ability to discuss your research with others. If you cannot do this, it is very likely that you aren't really sure what you are researching, which will make your research process very frustrating. It is not unusual for students to think that they have a grasp on their research topics, only to find that their finished papers wander from one subject to another. Instead of planning to do a paper on red tide, focus your search topic into something more specific, such as "how to inhibit the growth of red tide." You might ask the question, "What is the effect of volcanic eruptions on global temperatures?" instead of collecting all kinds of general information about volcanoes.

Note Taking

An easily overlooked skill that must be developed during the research process is the ability to document your research results as you progress through a literature search. It is extremely important to take good notes and to remember to write down accurate *citations* for the resources that you consult during your research. Many students proceed through their research by making haphazard photocopies of articles or book chapters, jotting down notes in margins and entering sketchy citations in their notebooks. Although it seems like more work at the time, it will save you a tremendous amount of time and angst if you learn to organize your research notes as you work. When you are ready to write your paper, good notes will allow you to proceed with ease through the writing process. Accurate citations in your notes and *bibliography* will enhance the credibility of your work.

One method that has worked well for many students is to use three-by-five index cards to take notes and keep track of their resources. Every time a new book, article, *government document*, Web page, or other information source is consulted, you should use a new index card and write down the complete citation for the resource.

When citing a book, you should be sure to include the complete name of the author or editor. Write down complete names for multiple authors if applicable. You should also write down the full and complete title of the book, the publisher, the place of publication, and the copyright date. Finally, write down the *call number* for the book so that you can locate it again easily if you need to check on something.

If you are citing a journal article write down the author (or authors) of the article, the complete title of the article, the complete title of the journal, the volume, issue number and date of publication for the journal, and the complete pagination for the article. If journals in your library are kept in call number order, jot down the call number on your resource card. Electronic journal citations will include all the information listed previously and the source where you obtained the *full-text* material—either the complete Web address (the *Universal Resource Locator* or *URL*) or the *database* name. You will also need to know the date that you accessed the article on-line.

Citations for *Web sites* include the author or sponsor of the Web site, the title listed on the Web site page, the complete URL for the site, the date the site was last updated or *copyrighted*, and the date that you accessed the site.

Government document resources will require the complete name of the agency that issued the report as well as the complete name of any individual authors who might be given credit for the work. You will also need to write down the full title of the document, the place of publication, the publisher (which is often the government agency or the *Government Printing Office*), and the date of publication. Be sure to include in your notes any report or document numbers that are listed in the document. It is also useful to write down the Superintendent of Documents (SuDocs) classification number for the document. This will help you to locate the document again should you need to check on any of the information. Finally, if you accessed the full-text of the document on-line, you will need to write down the complete URL for the Web site where you located the material and, if you obtained the full-text from a commercial database, you will need to cite the database name as well. Don't forget to write down the date that you accessed the report on-line. In chapter 10, we will discuss in detail how to locate and use government information.

Remember, this method works best if you use a separate card for each resource. Keep all these resource citation cards together. When you are ready to compile your bibliography and notes, you can organize all the cards into the proper order and simply start writing! Figure 2.2 shows what a resource card might look like for a journal article.

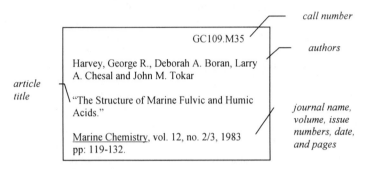

Figure 2.2. Note card illustrating the information needed to cite a journal article.

Once you have the resource cards completed you will want to begin taking notes. Three-by-five cards also work well for note taking. As you read through a book or article and find a bit of information that you decide is an important or key concept, write a paraphrased version of that information on your note cards. If you think you will be using an exact quote from the author's work, be sure to write on the card that it is an exact quotation and then carefully copy down the quote. At the top of the card, write a brief citation that will allow you to relocate accurately the resource that you used. Be absolutely sure to include the page where you located the information you are documenting on the card. Refer to figure 2.3 for an example of how to write up a note card.

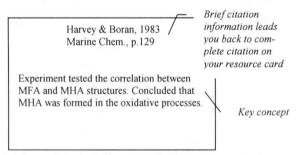

Figure 2.3. Note card illustrating a key concept with its corresponding resource information.

Continue to use the three-by-five cards to take additional notes. Although it may seem like a lot of cards, it is very useful to limit the information on each card to *one concept*. If you use this method, you can organize all your final note cards into one stack by putting similar concepts (from different resources) together. Working through your organized stack of note cards will allow you to maintain your focus while writing your paper without having to jump around in your notes.

If you have access to a laptop computer, you can use an electronic version of the method described previously. Keep a file that contains the complete information for citations to all the resources. In another file, jot down key concepts and quotations you have found. The concepts can then be cut and pasted into an organized set of notes. As you cut and paste using the laptop method, be very careful not to separate each note from its corresponding citation information. You still want to be able to give credit to the original authors accurately if you use their material in your paper.

Avoiding Plagiarism

As mentioned in the preceding section, it is extremely important that you give credit to an author if you refer to his or her work in your own paper. Computers and the *Internet* have, unfortunately, made it very easy for students to claim another person's work as their own. Not only is this ethically improper, it is illegal!

Most people know that credit must be given for direct quotations, but they may not be aware that there are many additional instances when you must give credit to an author. How can you tell if the information that you are using should be cited in a *footnote* or a bibliography? Factual information that is generally known or has been written by many different authors normally doesn't have to be cited because it is considered general knowledge. If you are citing your own opinions or the results of your own research, you do not have to provide a footnote. If you are referring to another author's opinions, insights, or research, however, you must document the original source of your material. Textual material is not the only thing that must be credited. Web pages, illustrations, figures, and computer software are among the items that must be given appropriate credit if they are used in your paper. If you have any doubt in your mind about whether or not to document your source, chances are you probably should. Finally, remember that giving credit to another author in your paper may not be enough to satisfy the *fair use* copyright guidelines. If you are using a substantial amount of another person's work, you must obtain their written permission prior to including it in your own research. This is especially important if you plan on publishing your research.

Style Manuals

Style manuals are used to provide guidance and consistency in writing and in citing resources within a paper. There are many different citation styles available. The APA (American Psychological Association) style is often used in scientific papers, but you may be asked to use a different citation style such as the *Chicago* style, the ACS (American Chemical Society) style, or the CBE (Council of Biology Editors) style.

Style manuals are normally consulted for specific guidelines on how to cite a footnote or bibliographic *reference*. These manuals will tell you, in detail, where to place commas, which words to capitalize, and in what order to place each element of the citation. Examples for books and journals are fairly straightforward. Citing technical reports and government documents can be a bit trickier, but examples are given.

The citations that students usually find most difficult to cite are those that refer to information they obtained electronically. Once a student realizes that electronic format differs from the citation style for print materials, they have a tendency to go overboard with their electronic citations. Suppose you use an electronic indexing database such as *Ecology Abstracts* or *GeoRef.* From these databases you locate several citations that look interesting so you check to see if your library subscribes to the journal in question. You find that the library has subscribed to the printed version of this journal for many years. You go to the periodical *stacks* and read the article from the bound journal. After reading the article, you decide that this resource has information that you would like to include in your paper. At this point, many students think that they must cite this resource using an electronic format since they located the citation on-line. This is not the case. The electronic database was simply a tool that you used to locate the citation. The content (full-text) of the article was available to you in its original printed format.

So when do you need to use the electronic style format? This is used when you obtain the actual content, the full-text of the article or book, on-line. Let's look at another scenario. Suppose you are reading an interesting article. In its bibliography, there is a reference to an article that appeared in the journal *Science* in 1888. You decide that you would like to read this article, but when you check your library holdings you find out that printed copies of *Science* only go back to the year 1900. You ask the reference librarian for suggestions and discover the older issues of *Science* are available through a database called *JSTOR*. Accessing *JSTOR*, you locate the 1888 article and read it as it is displayed on the computer monitor or you print a copy of the text to read later. In either case, since you used an electronic resource to obtain the full-text of the article, you should cite the resource in an appropriate format for an electronic resource. Figure 2.4 illustrates, using the *Chicago* style, the proper method for citing an article that was retrieved electronically.

Conn, H.W. "Bacteriology in our Medical Schools."
Science 11 (16 March 1888): 123-126.
http://www.jstor.org/ (6 December 2001).

**Figure 2.4. *Chicago* citation format for a journal
article retrieved electronically.**

The sample research strategy provided at the end of this book is included to help clarify how you need to cite various types of resources. If you are still having difficulty figuring out how to cite a resource correctly, check with the reference librarian for guidance.

Many students do not use care in compiling their references, and the results can be sloppy and unprofessional. As you progress to the point where you are publishing the results of your research, it will be necessary to be very precise in your citing style, so it is a good idea to develop the proper habits early on. Accurate citations allow others to

follow your research ideas and to locate your original sources of information.

Style manuals also provide guidelines for the formatting of an entire paper. The manual will tell you how to type up your title page,
what sections should be included, how to cite references within the text
of your paper, and how to insert tables and illustrations. It may also
give grammatical suggestions as well as ideas for improving your writing style.

Your professor or an editor can tell you what citation style you are
expected to use. It is useful to know what style you will be using before
you begin your research. You can then write down your resource citations in the proper style on your note cards or include them on your
computer file. Not only will this ensure that you have all the information that you need, but it will also save you time later when you type
your paper and compile your bibliography. If you have an extensive list
of references, you may want to invest in commercial database software.
These programs help to organize your bibliographic citations, and they
offer several automatic citation style formatting options such as APA or
Chicago.

Summary

The key to getting off to a good start in your research is to organize
your thoughts and focus on your topic *prior* to starting your research.
Make sure that your research topic is appropriate for the assigned
length of your paper and that the topic is not too general. If the topic is
quite broad, you may be able to focus your research by concentrating
on a particular population or geographic area, or by limiting your research to a particular time frame. You should be able to summarize
your research topic into a single sentence or question.

Once you have a good feel for your topic, do some preliminary
research to be sure that you will have access to the materials that you
will need and that you can obtain the resources prior to your writing
deadline.

As you begin your research, organize your note taking. Keep separate cards or files on your laptop that will allow you to collect similar
concepts into a final, cohesive order. Write down the *complete* citation
for every resource that you consult, even if that resource is not used in
the final version. It is better to have the information, and not need it,

than to try and retrace your research from incomplete notes. Find out what citation style your professor or editor prefers so that you can be sure to obtain all the information that you will need to complete your bibliography.

When you begin to write your paper, take great care that you give appropriate credit to another author's research if you include their insights in your paper. When compiling your footnotes and bibliography, pay close attention to the format listed in the style manual that you elected or were instructed to use.

Chapter 3

The First Step: Dictionaries, Encyclopedias, and Other Ready-Reference Resources

Often a student, who has reached the college or university level of their education, no longer sees any relevance in the most basic library reference materials, namely dictionaries and encyclopedias. As instructors, however, we see on a daily basis how many students may still benefit greatly from using these resources. Science students will be expected not only to perform experiments and observations but also to record and distribute the knowledge that they learn, both in their academic careers and beyond. This requires excellent written and verbal communication skills. Without these skills, many scientific advances would be lost.

Dictionaries

In addition to their primary goal of providing the meaning or definition of a word, dictionaries help us with correct spelling, word usage, pronunciation, syllabication, and grammar. Still think you don't need a dictionary? Dictionaries may also provide major place names such as rivers, mountains, and other geographical areas, as well as biographical information, synonyms, antonyms, and abbreviations.

The number of word entries usually determines the scope of a dictionary. An *unabridged dictionary* would include as complete coverage as possible for all the words of a given language. In the English language, those dictionaries with more than 265,000 entries are considered unabridged. Unabridged dictionaries describe obsolete and unusual words as well as current expressions. An *abridged* or shortened dictionary contains a selected range of entries.

There are four basic types of dictionaries. General dictionaries are designed to define words to meet the reading needs of nonspecialists—words that are used in everyday life. Subject-specialized dictionaries list and define terms or phrases that are frequently used in a particular field of study. The science student may find these science-oriented dictionaries most useful, especially if they contain supplementary information. Linguistic or historical dictionaries treat language like a science and trace the etymological origins of words and word usage. The *Oxford English Dictionary* is one of the best examples of this type of resource. Foreign language dictionaries translate from one language to another, rather than define the words of one language. A *polyglot dictionary* is one that translates a word into several languages. These too may be extremely helpful to the scientist who reads and writes materials for an international science audience.

Dictionaries usually define their coverage and scope in the pages that preface the main body of the work. Here, one may also find lists of abbreviations used in the work, a list of contributors, and other information that will make it easier to use the dictionary.

Hidden Treasure: Supplementary Material

Many general and subject-specific dictionaries provide information other than meaning, pronunciation, or the illustrations that may accompany a definition. These special features often appear in an appendix rather than in the main body of the dictionary.

General science dictionaries may contain the widest diversity of information that enhances the usefulness of the resource. The following is a selected list of the types of information that may be located in the supplementary section of a scientific dictionary:

Classification of the animal kingdom
Classification of the plant kingdom
Geological time charts
List of constellations
Properties of the sun, planets, and their satellites
Periodic table
Properties of elements, subatomic particles
Symbols used in electronics and mathematics
Units of measurement and conversion factors
Chronologies of discoveries and inventions

The more specialized the dictionary, the more detailed the information may be. Physical science dictionaries may provide much more detail about measurement systems and conversion factors such as ancient, Imperial, metric, and the metric-based International System of Units. Lengthier charts of physical and mathematical constants, abbreviations and symbols in those fields, and related equations are presented. Detailed isotope data, chemical densities, mineral properties charts, and a great deal more may also be found.

In addition to some of the information mentioned previously, a chemical dictionary may delve into the interesting origin or history of chemical terms. Did you know that cobalt came from the German word for "goblin," and that separating nickel and cobalt ore was so difficult that miners presumed that the devil or some evil spirit was interfering?[1] A history of chemistry, brief biographies of notable chemists, information on vitamins, and important organic ring structures are just some of the treasures that one may find in these specialized resources.

For the biologists, subject-specialized dictionaries may include data on biological fluids and chemicals, nutrition information, plant and animal classifications, and a chronology of significant events. People in the field of biology may also be included along with names of Nobel Prize winners.

Biological dictionaries may also include some medical data, but specialized medical dictionaries distinguish themselves by including an appendix on basic first aid, an atlas of human anatomy, addresses for medical societies and health organizations, lists of common medical procedures, medicinal measurements, nutritional and vitamin information, and even health tips for travelers.

Evaluating Dictionaries

In chapter 11, we will discuss at length the methods for evaluating sources of information. Although the dictionary is viewed as a quick answer resource, one should make an effort to ensure that the source is reliable. Begin by checking out the publication date of the resource. In some cases timeliness is not an issue but, in others, it may be critical.

Next, try reading through a few definitions. Are they clear and complete or do they simply refer you back to other terms? The reputation of the publisher may also be used to help determine the value of a dictionary. Finally, check to see if the dictionary includes useful supplemental information such as that listed previously.

It is important to remember that no single kind of dictionary is going to be sufficient all the time. Language is not static. Each dictionary will have its strengths and perhaps a few weaknesses. It is usually a good idea to consult with the preface and explanatory notes at the beginning of the dictionary to use it effectively.

The Encyclopedia

Encyclopedias expand on the dictionary entry by offering information in the form of articles of varying length on many subjects. The entries are usually organized alphabetically. An encyclopedia may be arranged as a single volume, similar to an expanded dictionary, or as a multi-volume set. When using a multivolume set, it is extremely important to use the companion index to find the subject you need. Good information does not always appear exactly where you expect to find it! Many encyclopedias include a bibliography or list of suggested readings for each entry. As with dictionaries, you should try to familiarize yourself with the preface and introductory materials before using the encyclopedia.

In general, the purpose of an encyclopedia is to provide an overview of a topic, background information for either the expert or the layperson, to provide brief answers to questions, and/or to provide bibliographies, which lead the researcher to additional material. In these ways, the encyclopedia acts as a starting point for further research, not the end of the research inquiry. As the encyclopedia represents accumulated knowledge, it is not reasonable to expect an encyclopedia to provide cutting edge news. The researcher should of course rely on professional conferences and journal materials for this information.

Evaluating Encyclopedias

There are two basic types of encyclopedias: general encyclopedias that condense and organize the knowledge from many subject fields and subject-specialized encyclopedias that limit their coverage to a particular field of study. You may find that the general encyclopedia does a better job of explaining a scientific principle than a specialized encyclopedia, particularly if you are looking for background information in a field that is unfamiliar to you. It depends on the need of the user. For this reason, the following issues should be examined.

Intended Audience

The intended audience of a general encyclopedia may vary, but usually it is not directed toward a subject specialist. Some publishers target a younger audience, while others may market themselves to the educated layperson. This is not always an indication of overall quality or usefulness, but it may guide you in selecting an appropriate resource. The specialized encyclopedia normally assumes that the reader is somewhat familiar with the language of the subject area presented.

Authorship

Both general and subject-specialized encyclopedias may identify the author of the article. This element of information may be used in evaluating the reliability of the article. Sometimes the encyclopedia will also provide brief biographical sketches of the author. If the article was written by someone whose credentials can be verified through other biographical sources, you may use the information with a greater degree of confidence. You may also be able to determine if the author has a strong bias that has influenced his or her presentation of the material.

Some encyclopedias identify the authors by using their initials rather than by spelling out the full names at the end of each entry. In this case, the resource usually contains a separate list of contributors where the full name can be identified.

Length

Length is not always an indication of quality, and lengthy articles may be found in both general and specialized encyclopedias, particularly multivolume sets. Since specialized resources limit their coverage to begin with, you may expect greater depth on the topics that they do include.

Bibliographies

A mark of a good article is the inclusion of a bibliography, i.e., the list of references used in the preparation of the article. You may also find a list of suggested readings to enhance the encyclopedia information. These lists can be extremely helpful in identifying additional sources.

These citations may also indicate if a balanced approach was taken in the presentation of the topic.

Timeliness

In order to produce new editions of a publication on a timely basis, it is often impractical to revise every entry in the encyclopedia each time a new edition is published. The user must pay attention not only to the publication date of the encyclopedia but also must look for other indications of how recently an article may have been updated. If you are somewhat familiar with the topic that you are investigating, you may observe the absence or inclusion of more recent advances in the field. Even if the encyclopedia was published within the past few years, some of the articles may be exact reprints of articles that were issued in previous editions. Does the article discuss any new developments that may have taken place within the last few years?

If you are not at all familiar with the subject area, you may have to test for timeliness by examining the dates of any references or suggested readings provided. If these dates are substantially older than the publication date of the encyclopedia, then the article may not have been updated recently. Depending on the subject matter, this may or may not be a problem to the user.

Illustrations

Illustrations such as photographs, maps, charts, tables, and other diagrams should be appropriate to the topic and enhance the text rather than add clutter. All illustrations should be clearly identified and explained. Illustrations won't make or break an article, but the relief from continuous text and the clarity that an illustration may provide are often welcome.

Ready-Reference Sources

Ready-reference is a librarian's term so let's define what we mean by the "ready" in that phrase. Ready-reference resources provide concise answers to short or fairly simple questions quickly, and readily. These sources are frequently published on an annual basis. If so, they have the advantage of containing more current data and statistics as well as compiling historical information. Encyclopedias and dictionaries may be used for some ready-reference questions; but as they may not be published or revised every year, we may not always be able to rely on their timeliness.

There are several different types of resources that fall under this category: *almanacs, yearbooks, handbooks, manuals, directories*, and geographical resources.

Almanacs and Yearbooks

An almanac is a compendium of useful current and retrospective data and statistics. The scope of the almanac may be general with information on countries, events, persons, or it may be subject-oriented. A yearbook or *annual* is a collection of data and statistics for a given year. The purpose of a yearbook is to record recent activities or to update previously published material.

Handbooks and Manuals

Handbooks and manuals are closely related to one another. Both are published on an as-needed basis rather than annually. A handbook compiles basic facts on one central subject area. It is particularly useful to have scientific handbooks readily available to you. These resources provide a good place to check for many of those specific details that you learned once but can't quite remember. In various scientific handbooks, you will find formulas, conversion tables, wind speeds for the Beaufort scale, geological time scales, basic taxonomic classifications, and much, much more. *The CRC Handbook of Chemistry and Physics* is a well-known example of a scientific handbook that is considered an essential purchase for many laboratories.

A manual usually refers to a how-to type of publication but, again, may also be published in more of a handbook format. Good examples of a manual are the style manuals that provide information on how to set up a term paper or manuscript. These books, such as the *Publication Manual of the American Psychological Association* or the *Chicago Manual of Style*, give very specific details on how to cite references and format a paper.

Directories

Need the address and some basic information about organizations that provide funding for scientific *grants*? Or, do you need to locate the phone number for an out-of-state business that manufactures laboratory supplies? These types of questions can be answered by consulting an electronic or printed directory. Directories are lists of persons, places, or organizations, usually arranged alphabetically for ease of use. Remember though that the type of information that you find in a directory changes very quickly, so it is particularly important that you use the most recent version available.

Geographical Sources

Geographical sources include *maps, atlases, gazetteers,* and even travel guides. You never know to what part of the world your science will take you! A map is a representation of certain features of the earth, or part of the earth or other body on a flat or spherical surface; the latter is generally called a globe. An atlas is a bound collection of maps with accompanying text and index. The gazetteer is a geographical dictionary that normally does not include any maps. Gazetteers are helpful with spelling of place names and physical descriptions such as latitude and longitude. They are also a useful place to begin if you are completely unfamiliar with a place name. By using a gazetteer, you can quickly locate the general part of the world where a place is located. From there, you can go to a map to pinpoint the exact location visually. Travel guides highlight interesting and frequently necessary information about specific regions or countries.

When using geographical sources, you should check publication dates to ensure timeliness of information. Geography is definitely not a static science as place names and territorial boundaries are changing regularly. Also the scale and *projection* (or distortion) should also be considered depending on the kind of work you are doing. Lastly, most geography tools utilize symbols to represent certain features and will provide a key to the meaning of these symbols.

Summary

You are never too old or too smart to need a few good dictionaries. In addition to providing definitions, spelling, word usage, pronunciation, and grammar, many dictionaries provide valuable supplemental material that can help you to quickly answer basic questions.

The encyclopedia is an excellent place to start looking for background information or help with terminology. The encyclopedia article may provide additional sources of information and names of authorities in the field through the signed articles. Remember to start with the index volume of a multivolume encyclopedia to locate your topic. The disadvantage of an encyclopedia is the lack of timeliness, as the publishers do not revise every entry each time the encyclopedia is published. There may also be inaccuracies despite the most careful editing. As with any resource, you may notice some bias on the part of the author.

Ready-reference materials such as almanacs, yearbooks, handbooks, manuals, and directories are used to provide quick answers to brief factual questions.

Note

1. Richard J. Lewis, Sr., *Hawley's Condensed Chemical Dictionary*, 13th ed. (New York: John Wiley & Sons, 1997), 1211.

Chapter 4

Getting Down to Business

The first step in research may sound simple, but it is crucial. Identify the broad subject area within which your topic falls. Most of the tools and methods for finding information are based on subject content. You probably don't want to use a resource that concentrates on criminal justice research if you are looking for material on organic chemistry. To locate the most relevant information you must select a resource that covers your area of research.

Selecting the Right Tools

In chapter 3, we discussed how subject-specialized science encyclopedias and dictionaries can serve as wonderful sources for finding answers to brief questions or for locating background information. Other types of science-related directories, handbooks, manuals, and almanacs may often provide more pertinent data than multipurpose or general sources.

Periodical indexes are tools for locating *journal* articles. Many periodical indexes are subject-specific to a particular discipline. Obviously, you don't want to use a political science index if you are interested in finding articles on volcanoes. How then do you determine what resources are available in your area of interest?

A great place to start is by asking the reference librarian. He or she is most familiar with the reference collection and may give you suggestions for the most appropriate science resources in your library.

You may also search the library catalog. Libraries using *Library of Congress Subject Headings* assign *subdivisions* such as periodicals, indexes, or handbooks to broad science subject areas. For example, if you wanted to locate periodical indexes in the field of chemistry you might go to your on-line catalog and look up the subject:

    chemistry—periodicals—indexes

Additional information about using subject headings is discussed later in this chapter.

Getting to Know the Tools

The best way to learn about any resource is to examine the preface and introductory material. These pages are found in the front of each volume. This information is usually present in an electronic database as well, but it may be more difficult to locate. Normally, this section of the database is identified by a button or link labeled "information" or "about the database." The on-line help screens may also provide useful information about the resource.

When selecting a resource that you have never used, there are some important features to examine and questions you should consider.

Dates of Coverage

What months or years are covered in the specific volume or electronic resource you are using? Has this source been updated on a regular or reasonable basis? If you require historical information, is this sufficiently covered by this source or does it only cover recent materials?

Printed resources are normally not updated except when a new edition is published or when a supplemental yearbook is issued. Materials that are printed in a loose-leaf format may be updated more regularly. In those cases, updates are sent to the library and the outdated sheets are replaced with new sheets containing more current information. Also, as mentioned in the chapter on encyclopedias (chapter 3), just because a new edition is published, the sections that you are using may, or may not, have been updated with the new edition.

Electronic resources vary widely in the reliability of their updates. Some are updated daily or even hourly. Others may be updated on a quarterly or yearly basis. Unfortunately, just because a resource is available electronically is no guarantee that it is current. In rare cases, an electronic resource may not be as up-to-date as its printed counterpart. The policy on updates should be available in the information about the resource.

Type of Material Covered

Will you be able to locate the *kind* of material that you need? Many resources include different types of material. Encyclopedias may incorporate information from books and magazines. Handbooks may utilize government-produced statistics and research. Many indexes cover selected books, dissertations, government documents, technical reports, and conference proceedings, as well as periodicals, if the publishers feel the material is relevant to their subject areas.

Intended Audience

What is the reading level of the text? Some reference resources are intended for high school students. Others are written in a very technical language that only graduate students, faculty, or professionals in that field of study may understand. Again, the introductory material should give you a good idea of the intended audience level. Ideally, you will want to locate resources that are research based but not so technical that you have great difficulty reading and comprehending the author's points.

Format

Format may vary from one publication to another. Many people assume that all reference books are arranged alphabetically. In some cases, the book may be arranged chronologically, by topic, by taxonomic classification, or in some other manner. The introductory material will explain how the resource is organized and indexed, and will often include a sample citation to illustrate how the entries are formatted. Any abbreviations that are used within the resource are also normally defined in the introductory material.

Defining Your Search Terms: The Use of Controlled Vocabulary

Once you have selected some appropriate subject-specialized research tools, you will want to use them as efficiently as possible. The words and phrases that you choose to define your research topic may greatly influence how much information you retrieve. Learning to use a *controlled vocabulary* can significantly improve the *relevancy* of the materials that you locate in your search.

A controlled vocabulary is a compilation of subject terms or phrases, which have been selected and authorized for use in describing materials before entry into a catalog or database. This compilation of these terms is also known as a *thesaurus* and, for our purposes, we will use the terms controlled vocabulary and thesaurus interchangeably.

Unfortunately, there is no one standard thesaurus that can be used with all indexes. Some publishers will utilize the thesaurus that is published by the Library of Congress called *Library of Congress Subject Headings (LCSH)*. Others, especially those that prepare and publish materials for specific subject disciplines, may find it more advantageous to develop their own authorized list of terms, which better defines the vernacular of that profession or field of study.

Indexers and catalogers are people who are responsible for describing books, documents, journal articles, and other materials by their subject content. They rely almost exclusively on the appropriate controlled vocabulary when assigning subject headings to provide the consistency needed.

Let's demonstrate how this all works to our advantage as researchers. You have come to the library to find some materials on the drugs that are being used to treat the disease AIDS. You want to look up the disease, but do you look for "aids," "AIDS," "acquired immune deficiency syndrome," "acquired immunodeficiency syndrome," "HIV infections," "viruses," or "immunosuppression"? There are so many possibilities! How do you know where to start searching?

In order to provide some order to this potential chaos, a researcher may consult a controlled vocabulary or thesaurus to discover what terms have been authorized for use in a particular resource. If the resource that you are using has chosen to use "acquired immunodeficiency syndrome" as its official way of describing the disease, indexers will assign any book, *dissertation*, document, or article treating the subject that *exact* subject heading. Even if the title of the book or article refers to the term "AIDS," the indexers will still look to their official terminology and use "acquired immunodeficiency syndrome" as the subject heading for the material. If you know what official subject heading an indexer has selected, you are relieved of the burden of thinking up all the possible words and phrases that may have been used to describe the subject you are researching. The use of a thesaurus ensures a measure of consistency in a library collection or information database. A thesaurus also allows you, the information seeker, to select vocabulary terms that will make your search more effective and efficient.

How to Select and Use a Thesaurus

In most instances, the catalog or database identifies the thesaurus used in the organization of the resource. In a print source, this information is usually available in the first few pages of each volume where the mission and scope of the resource are defined. In electronic resources, the thesaurus may be identified in the information pages, or there may be a link to the on-line version of the thesaurus if it is available. If you aren't sure which thesaurus to use for a specific index, ask a reference librarian for assistance.

A thesaurus usually organizes authorized terms or phrases alphabetically like a dictionary, but the similarity ends there. A dictionary defines words and helps with pronunciation. This is not the main mission of a thesaurus, although an occasional *scope note* may describe how that term is being applied in context. The purpose of a controlled vocabulary is to identify similar terms and describe the relationships between them.

Under an authorized term or phrase, you will normally find one or more related terms. In some publications, the relationship between the *main entry* term and the related word is further defined through the use of codes. In figure 4.1, words listed with the codes BT, RT, NT are broader, related, or narrower in scope than the main entry term.

Broader terms (BT) and *related terms (RT)* can be used to expand your search. In this case, if a search of "remote sensing" did not produce satisfactory results, you could try looking for information on the subjects "aerial photogrammetry" or "space optics." Conversely, if your search of "remote sensing" retrieved too much information, you can look at the *narrower terms (NT)* for ideas on how to limit your

Remote Sensing *(May Subd Geog)*
   [G70.39-G70.6 (General)]
        Here are entered works on the theory or methodology of collecting and gathering images of distant objects or property. Works on handling, maintaining, and indexing remote-sensing images in unbound collections are entered under Remote-sensing images.
      UF   Remote-sensing imagery
            Remote sending systems
              [Former heading]
            Remote terrain sensing
            Sensing, Remote
            Terrain sensing, Remote
      BT   Aerial photogrammetry
            Aerospace telemetry
            Detectors
      RT   Space optics
      SA   *subdivision* Remote sensing *under topical headings*
      NT   Aerial photography
            Artificial satellites in remote sensing
            Electronic surveillance
            Microwave remote sensing
            Multispectral photography
            Radar
            Thermography
  — **Equipment and supplies**
      NT   Fire Detectors
            Imaging systems
            Infrared horizon sensors
            Scanning systems
            Star trackers
            Sun trackers
  — Mathematics
     [G70.4]

Figure 4.1. Example from *Library of Congress Subject Headings* 2000, (23rd edition), p. 5058. Copyright © 2000 the Library of Congress.

search, e.g., use "electronic surveillance" or "microwave remote sensing." Another method used to focus a search is to apply subdivisions to the main subject heading. In figure 4.1, the subdivisions are identified by dashes under the main heading of "remote sensing." So, if you were interested in finding materials on equipment used for remote sensing, you would look up the subject "remote sensing—equipment and supplies." The exact meaning of each code is explained in the introductory section of the thesaurus.

In order to standardize similar terms under a limited number of subject headings, unauthorized terms may be listed in the thesauri as well. There are many common expressions that are not assigned (used) as subject headings. In these cases, the thesaurus will list the unauthorized term and direct you to a related term that is used in its stead. Many thesauri display the symbols *"USE"* and *"UF" (use for)* to lead you to the appropriate subject headings. In figure 4.1, you can see that the term "remote sensing imagery" is not an authorized subject heading since it has a "used for" code. The catalogers decided to use the term "remote sensing" in its place.

So what happens if you use an "unauthorized" subject heading? If you are using a printed resource you simply won't find an entry under the unauthorized heading. If you perform a subject search on an unauthorized heading in an electronic resource, your search will not retrieve any *records*, or the database software may provide a link to the authorized term.

The Importance of Controlled Vocabulary

You should use a controlled vocabulary tool or thesaurus for the same reason you might use the index to a book. You don't want to thumb through every page of the book when you only need a specific part. For example, when we use a multivolume encyclopedia, we look at the index because we are not sure which volume may contain the information we need. It saves us lots of time and energy. Thesauri can save time and energy as well, especially when searching electronically since there is so much information available. Many of the electronic systems allow *keyword* searching, which is a very broad, fuzzy sort of way to search. A keyword search will look through the entire database for a particular word or phrase (see chapter 5). Under certain conditions, this is not necessarily a "bad" search technique, but it can be a bit like trawling with a huge net over a wide area. You are going to pull in lots of fish but probably a bunch of old fishing line, discarded boots and sneakers, plastic milk cartons—in other words, JUNK, as well. You may be overwhelmed (or underwhelmed depending on your expectations) by it all.

Remember, catalogers have assigned subject headings to these items only if they represent the *main concepts* in the material. If you search using these subject headings, the results of your search should be much more relevant to your topic.

Thesauri may also lead you to consider subject terms or phrases you may not have considered using on your own. In the next chapter we will illustrate how the subject headings you select can be combined to formulate an efficient, focused *search strategy*.

Summary

A critical step in research is the selection of appropriate subject-specialized information tools and resources. You can locate the most useful tools by:

- Asking a reference librarian.
- Using the library catalog.
- Understanding the scope and content of the resources you have selected.

A thesaurus can be an extremely useful tool in library research. It may:

- Help to define a topic more clearly.
- Help to expand or limit a search as needed.
- Identify useful terms and phrases to use in searching.
- Suggest terms or phrases that would not otherwise have been considered.
- Improve the relevancy of searching through the use of subject headings.

Chapter 5

Composing Your Search Strategy

In previous chapters, some general techniques were discussed that relate to selecting and refining a research topic. Some information requests are very simple. We can look up an idea in a thesaurus, grab one subject heading, and use that heading to find information in various resources. Some questions, however, are not that simple. They may involve two or more different topics or factors, and each needs to be represented in a search statement to retrieve relevant information.

Printed sources are very straightforward in this regard. Subject headings appear on a printed page that is easy to read. The addition of subdivisions to a main subject heading combines two topics for us. That is, the main topic is represented by the subject heading and is further refined by the subdivisions or secondary subjects.

Electronic databases are more complex, but at the same time, more flexible. In most cases, the computer software does not intuitively "understand" what is needed. For this reason, it is important to prepare a *search strategy*. A well-constructed search strategy combines standardized search commands with appropriate subject terms and phrases in a format that can be processed by the computer software.

Consider Your Source

Before beginning a detailed discussion about search strategies, it is necessary to emphasize the importance of selecting an appropriate resource in which to search. Making the effort to prepare a great search strategy is a waste of time if you start looking in the wrong place! Knowing something about the content and features of the resource, be it print or electronic, is extremely important. This philosophy will be stated many times in relation to different resources throughout this guide. If you are unsure about which resources to select, there are several methods you can use to help locate useful materials. Often the source will have descriptive information listed in the introduction. A librarian may have prepared a bibliography that describes available resources in your particular subject area. Of course, you should always feel free to ask the reference librarian for guidance.

Selecting Key Concepts

Novice searchers have a tendency to use too many terms when they search for information. Consider the topic "the effects of oil spills on sea otters." Many people would go to the computer and type:

the effects of oil spills on sea otters

While this form of query, called *natural language searching*, is easiest and becoming more common, it is still not available in many databases. In most cases, this kind of search statement will only serve to complicate the process. Many databases have *stop words* or words that are so commonly used that the computer can't reasonably search for them. Common stop words include, but may not be limited to:

a   an   be   for   is   of   on   the   to

In some databases, the searching software will simply ignore stop words and allow you to continue your search, sometimes with unpredictable results. Others will give an error message if stop words are included in the search strategy.

Returning to the topic, it should now be obvious that the words "the," "of," and "on" should be eliminated from the search statement since they are all stop words. This leaves the terms:

effects   oil spills   sea otters

On examination of the three remaining phrases, we might also consider eliminating the word "effects." Since most books or articles discussing oil spills *and* otters are probably addressing the issue of "effects" of the oil on the animals anyway, adding the word "effects" is redundant and may restrict the final search results too much. Therefore, the remaining key concepts are: "oil spills" and "sea otters." Once the key concepts are determined it is possible to begin working on selecting specific *search terms*.

Locating Search Terms

The next step in the search process is to select words or phrases that best describe the key concepts. Determine if there is a thesaurus associated with the resource you plan to use. The thesaurus may be located on-line as part of the electronic database or it may be available in a printed format. Remember to check with the reference librarian for help if you have difficulty locating an appropriate thesaurus. If a thesaurus specific to the database you plan to use is not available, it may be useful to examine a thesaurus from a different but subject-related resource. The related thesaurus may still provide useful suggestions for terms you may not have considered. Jot down any narrower, broader, or related terms that might be useful in your search. Refer back to chapter 4 for additional details on how to use a subject-specialized thesaurus.

Now let's return to our search on "the effects of oil spills on sea otters." We have decided that the key concepts are "oil spills" and "sea otters." After checking several thesauri, there are a number of possibilities for actual subject headings that will describe these concepts. Figure 5.1 illustrates some of these ideas. Of course, it is not necessary to use all the subject headings that are listed, but they may be useful later on if it is necessary to expand or narrow the search.

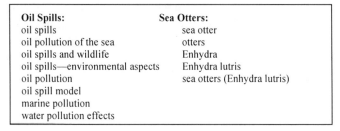

Oil Spills:	Sea Otters:
oil spills	sea otter
oil pollution of the sea	otters
oil spills and wildlife	Enhydra
oil spills—environmental aspects	Enhydra lutris
oil pollution	sea otters (Enhydra lutris)
oil spill model	
marine pollution	
water pollution effects	

Figure 5.1. Potential subject headings for the concepts: oil spills and sea otters.

Combining Search Terms

Now we have some specific subject headings with which to work:

> oil spills
> oil pollution
> Enhydra lutris
> sea otters

It's time to construct the actual search strategy -- the exact method used to instruct the computer to search for these topics. The first step will be to select appropriate *Boolean commands*.

What's a Boolean Command?

The concept of Boolean commands sounds a bit complicated, but Boolean commands are critical to effective on-line searching and, once you understand how they work, they are not terribly difficult to use. There are three Boolean commands: *AND*, *OR*, and *NOT* (which is sometimes displayed as AND NOT or BUT NOT)

For fans of set theory, the following concepts will probably be very easy to understand. If not, hang in there for an explanation of how all of this applies to library research. AND, OR, and NOT are used to combine search terms in a logical way so that the computer software understands what we want to retrieve from the database.

Using AND between two terms (or phrases) means that the computer software should retrieve only those records that contain *both* terms. If one term is included but the other is not, the record will not be of use and will not be included in the final search results. The AND command is used to *narrow* or *focus* a search by combining two or more terms representing *different* subjects. In the search on sea otters, the AND command should be used to combine the two main subjects: sea otters AND oil spills.   The goal is to retrieve only books or articles that contain both of these topics, so if the item discusses sea otters but it does not mention oil, we are not interested.

In contrast, the OR command is used to expand a search. When using OR to combine two or more *related terms*, the computer software is asked to retrieve records that have *at least one* of the terms listed. Some records may include all the terms listed in the OR search statement, but others may include just one of the terms. Remember to use OR only when combining *similar* terms. One way to help determine whether to use AND or OR is to return to the controlled vocabulary headings listed in the database thesaurus. Related terms listed under a heading may usually be combined using the OR command. In this search, "sea otters" and "Enhydra lutris" are synonymous terms, as they are both names used to describe the same animal; therefore, they would be combined with OR. "Oil spills" and "oil pollution" are also similar so they may also be combined using the OR command:

> sea otters OR Enhydra lutris
> oil spills OR oil pollution

Putting It All Together

It is now time to put all the search terms together in a logical manner:

> sea otters OR Enhydra lutris
> AND
> oil spills OR oil pollution

It is important to organize the search in an efficient and logical manner that makes sense to the computer database that you are using. Search systems may vary in the *order* that they perform Boolean commands. That is, some may perform AND functions first, then OR functions, or vice versa. Others perform each command in the order it appears in the search statement. This gets pretty confusing and can mess up the results of a search if the terms are not combined in the order intended. Let's look visually at a couple of possibilities. Here is the search statement:

> sea otters OR Enhydra lutris AND oil spills OR oil pollution

To get the proper final results, the system must perform the OR commands *first*, then combine the two resulting sets by using the AND Boolean command. In figure 5.2, the crosshatched area (intersection of sets) represents only those items that mention both otters and oil, but

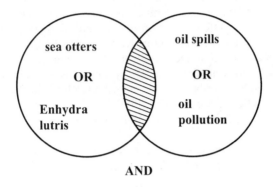

**Figure 5.2. Venn diagram illustrating desired search
results using proper Boolean techniques.**

the search system could perform the commands improperly. In the example shown in figure 5.3, you can see what would happen if the system processes the commands in the order in which they appear. The
illustration on the last line represents the final results. Since "oil pollution" was added into the equation at the end of the search using the
Boolean OR operator, the final result contains *all* the resources in both
of the final sets rather than the selected resources that appear in the
intersection of the final two sets. The end results include several types
of resources: those that discuss both sea otters (E. lutris) and oil spills,
some resources that discuss sea otters and oil pollution, and some resources that are about oil pollution but do not mention sea otters in any
way—either by common name or by scientific name. What can be done
to minimize this potential problem of order? Many search systems allow the use of parentheses to organize a search statement further.

sea otters OR Enhydra lutris AND oil spills OR oil pollution

now becomes:

(sea otters OR Enhydra lutris) AND (oil spills OR oil pollution)

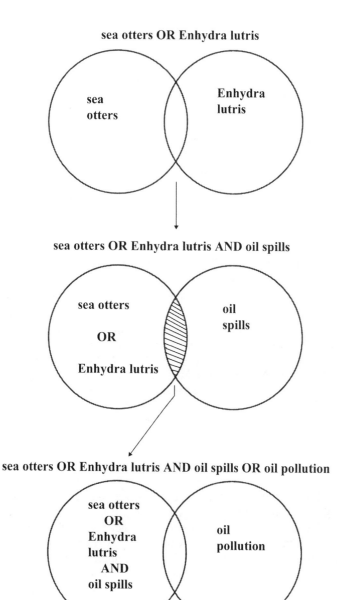

Figure 5.3. Venn diagram illustrating incorrect search retrieval resulting from the improper use of Boolean operators.

The parentheses tell the software in what order the search commands should be performed. Just like in algebraic statements, the software will look *inside* the sets of parentheses and perform those operations *first*.

The first combination of records will retrieve articles that include one or both terms "sea otters" or "Enhydra lutris." A second set of records will contain either or both of the terms "oil pollution" or "oil spills." Finally, the software should combine these two sets for a final result that contains information on both of the main subjects, otters and oil. Many search systems are being designed or modified in such a way as to eliminate the need for parentheses but, if in doubt, it is often best to try using parentheses.

Using the Boolean NOT Command

The Boolean NOT command, while still quite useful, is a technique that may require a bit more thought. Adding NOT followed by a term to a search statement removes any record containing that term from your results. This can be very useful in certain cases, particularly when trying to reduce false hits. *False hits* occur when the computer retrieves the search term you requested, but it is not in the context that you need. For instance, if you wanted to search for information on the disease AIDS, false hits or *false drops* may occur if the computer software retrieves records that contain the term "audiovisual aids." The NOT feature will work well here by using the search statement:

aids NOT audiovisual

This search strategy will remove all records that contain the phrase "audiovisual aids." Of course, you could also lose a relevant record if there happened to be an audiovisual aid that pertained to the disease AIDS. So, there is the possibility that you may restrict your search too much by using NOT and perhaps lose useful information as well.

As another example, suppose that you wanted to locate articles on the birds of Florida, but in your preliminary searching you found that many of the articles were devoting a lot of the text to birds of other southeastern states. You might consider searching:

(birds AND florida) NOT southeast

This strategy will eliminate the records that discuss birds in the southeastern portion of the United States (and in the southeastern section of Asia for that matter), but it will also eliminate records about birds in southeastern Florida. In addition, it will eliminate the articles that discuss the Southeast *including* Florida. If the record for a really great article mentions the word "southeast" *anywhere* in the record, that citation will be eliminated even though a significant portion of the article may contain really useful information on Florida birds. The best advice then is to use the NOT command as it is needed but always keep in mind exactly how it may affect your search results.

Field Searching versus Free-Text Searching

On-line resources have become so plentiful that locating information is normally not a problem. What has become difficult is locating good, relevant information—pinpointing the exact topic that you are researching. You may perform a search that retrieves more than 100 records, but are all of those records really useful, or do they skirt around the topic? How can you find a focused, relevant set of citations? Boolean searching is certainly a good first step, but there are ways to refine your search even more. *Field searching* may significantly improve the relevancy of your search results.

Whether searching in an on-line catalog of books or using an index to periodical articles, each record in the database is composed of a number of smaller sections called *fields*. In each record, the same field contains the same sort of information relating to the item being indexed. The search system associated with the database has the capability of searching many of these fields individually. Most on-line databases allow a *minimum* of four types of searches:

author
title
subject (sometimes called a *subject heading* or a *descriptor*)
keyword

In addition to the author, title, and subject fields, databases may have additional fields of information such as the publisher, the date of publication, a summary (*abstract*) of the item, or the language in which the material was written, all of which may be searchable. In field searching, the computer software is being asked to search for a specific word or phrase in the specified field only. For instance, a search for wolves using "wolf" as a keyword will pull up many items by authors named "Wolf," which probably will not be relevant to your work; a field search on "wolf" as a subject will eliminate these false drops. An efficient, experienced searcher will reduce or eliminate false hits by performing a subject field search on the exact subject heading that has been authorized for use in that particular database.

If you choose not to use one of the field searching options, you can perform a *free-text* or keyword search. In a free-text search, the computer's search system will look for the search term throughout the entire record. The record will be retrieved if the requested search term appears *anywhere* in any field in the record.

Field searching is normally the most efficient method of searching. The processing time is often shorter because the computer is searching only a small section of a record rather than the whole thing. Most important, a field search will usually produce the most relevant search results. As you recall from the discussion of controlled vocabulary in chapter 4, indexers assign subject headings to a particular record *only* if that subject is discussed in detail in the book or article being indexed.

The disadvantages of a free-text search may be, conversely, longer processing time. Most significant, it requires you to think of every possible word or phrase that might be used to describe your subject. It is a less selective way to search that may retrieve too many results and increase the number of false hits.

Why Would You Want to Perform a Keyword Search?

There are good reasons for performing a keyword, free-text search. In some cases, a topic is so new that it hasn't yet been assigned a subject heading. Imagine what fun indexers had when computers first appeared on the scene, long, long after library *classification systems* had been developed! Idiomatic expressions, acronyms, slogans, and other subject-specific jargon are frequently not listed as authorized subject headings, and yet everyone uses those words and phrases. There are also subjects that just have not been heavily researched anywhere, making it hard to locate information on that topic. In those cases, you may be willing to examine *any* material that mentions your topic, even if it isn't the main emphasis of the book or article.

If a thesaurus is not available for the database that you are using, you may begin your search by performing a keyword search on your topic. Even if you retrieve a large number of records, you can examine a few of them to see what subject headings were assigned to the records. Once you have determined some "authorized" subject headings, you can redo your search more efficiently using a subject field search. Your second search should be much more focused and relevant. In our earlier search about otters and oil spills, if you were unsure whether the database used scientific or common names as subject headings, you could perform a keyword search on:

sea otters OR Enhydra lutris

Examine the records that you retrieve to find out how the animal is described in the record, and then reconstruct your search strategy accordingly.

Examining Your Results

Once the software has processed your search request, you will be presented with the results of your search. First, examine the number of records that you have retrieved. If the search results equal zero, you obviously have to reexamine your research strategy. Check for errors in typing. A misspelled word will not retrieve results, unless the publisher misspelled it in the same way! Have you included stop words that might be causing errors? Have you used too many AND commands in your search strategy, overly restricting the search? Remember, every time you use an AND command, the terms on both sides of the AND *must* appear in the final results. Have you checked the database thesaurus for appropriate terms, or perhaps you need to free-text search? Lastly, are you are using an appropriate subject-related database?

If you have retrieved a few records but not enough to suit your needs and your search strategy looks okay, go to the thesaurus and look for broader terms that might help to expand the search. Look at one or two relevant records and the assigned subject headings to see if they might provide additional ideas for terms that might help to strengthen your research strategy.

Many people find that retrieving a small amount of information from a search is not as big a problem as retrieving too much information. If you are writing a thesis or dissertation or the "ultimate guide to whatever," you may need to locate every item that has ever been published on your topic. Most of us, however, are looking for a few good sources. How then do you refine your search to locate those few, extremely relevant records? As always, a good first step would be to go to the thesaurus to see if you can find narrower topics that will help to focus your search. You might try adding additional qualifiers to the original search to refine it further. Avoid using broad subject headings in databases that specialize in the area. For example, don't search for a word like "medicine" in a medical database such as *Medline*, "zoology" in the biological database *Zoological Record*, or "chemistry" in *Chemical Abstracts*.

In the next chapter, we will learn some advanced search techniques that will help you to refine your research even more.

Summary

Boolean searching is one of the most important techniques that can be used to focus a search. In order for the computer software to perform the search correctly, the searcher must clearly understand the function of each Boolean operator and how the computer software will interpret the command. Remember to place the Boolean commands in the proper order within the search statement, using parentheses as necessary. Boolean commands and their functions are listed below:

AND	combines different topics	normally focuses a search
OR	combines related topics	normally expands a search
NOT	removes a term from the search	normally focuses a search

Another way to improve your searching technique is to utilize field searching. There are several advantages to field searching:
- usually provides fewer, more relevant results
- usually provides fewer false hits
- may reduce computer processing time

In some instances, it may be necessary to expand a search through the use of free-text searching. Free-text searching searches for the specified term or phrase throughout the entire record. You may want to employ free-text searching if:

- the topic you are researching is not adequately described by authorized subject headings available
- the topic is too new to have been assigned a subject heading
- you know that your retrieval rate will be very low and you are interested in examining any book or article that even mentions your subject

So what happens when the search results are not what you expected? Use the following suggestions to troubleshoot your research strategies:

Zero hits:

- check typing and spelling
- check search strategy—did you use too many AND commands?
- did you use stop words?
- try checking the thesaurus for broader or related terms

Too few hits:

- consider adding related terms using the OR command
- try checking the thesaurus for broader terms

Too many hits:

- consider using narrower terms
- consider adding terms using the AND command

Chapter 6

Advanced On-line Searching

At this point, you may be asking yourself why you would need any more information on search techniques. It would be remiss if we didn't point out just a few more tips that can help you to develop strategies for successful searching. As pointed out in the section on beginning searching techniques in chapter 5, there may be differences in database search commands and functions but, if you understand the basic principles, you'll feel confident in your searching skills no matter which system you use. In this chapter, we will present some advanced search techniques and tips that are used in many of the more sophisticated search systems.

Truncation and Wildcards

In chapters 4 and 5, we discussed how to select appropriate search terms; but how are different forms of the same term or variant spellings of a particular word handled in a good research strategy? Consider again the example that was used in chapter 5 for the search on otters and oil spills. We decided that "pollution" might be a useful related term. Including other forms of the term "pollution" such as "pollute," "pollutes," "polluted," or "polluting" may help to expand the search. It can be difficult to anticipate every variation of a single word. Fortunately, the developers of most search systems anticipated this problem and defined methods to help solve this problem.

The common root to the previous words is "pollut-." A technique called *truncation* utilizes the root of a word and looks at all the variant forms of the word. To truncate is to shorten by cutting off letters at the end of the word. Truncation should not be confused with abbreviation, which is to shorten words by leaving out some of the letters, but not necessarily those on the end. When we want our search to include more than one form of a word, we type the root of the word ending with a *truncation symbol* that will represent the rest of the letter combinations. Different databases use a variety of symbols to indicate to the computer software that a word should be truncated. Various truncation symbols include: **?   *   +   ! $ #.** These symbols may represent none, one, or several characters at the end of a word. For example, "pollut*" may retrieve "pollute," "polluted," "pollutes," "polluting," or "pollution." So by using the truncation symbol, the software will search for one or more words even though only one search term was actually entered.

It is not uncommon in the scientific literature to find spelling variations depending on where the book or article was published. Examples of this kind of variation include the words "color" and "colour" or "behavior" and "behaviour." Sometimes it may be difficult to know exactly how a particular word or species name is spelled, simply because it is unfamiliar. Such is the case with proper or foreign nouns or with words that aren't readily verified in a common *abridged dictionary*.

Many of the same truncation symbols act as *wildcards* that may substitute for none, one, or multiple characters *within* a word. Suppose we want to search for the scientific name for sea otters, but we aren't positive of the spelling. Is it "enhidra," "enhiedra," or "enhydra"? We could try to verify the spelling or we can save a little time and use a wildcard. By entering "enh*dra" the system will retrieve all the variations of words beginning with "enh" and ending with "dra."

While this is an extremely helpful feature, if you truncate a word too much you may set yourself up for a lot of false hits. For example, if searching for forms of the word "pollution" with the truncation "poll*" you might also re-trieve "poll," "pollen," "pollinate," "polliwog," or "pollock." In addition to a much larger retrieval set, the relevance of the set can be compromised. Likewise with wildcards, try to anticipate if the substitution will be an effective way to search. If you are unsure whether the system you are using allows the use of truncation or wildcards, check the on-line help screens.

Proximity Searching

In chapter 5, we discussed search techniques that use field and free-text search-ing and how they may affect the quality of the search results. Remember, when you use field searching, you are restricting your terms or phrases to specifically assigned fields. In free-text searching, you are asking the search system to look for those terms in any searchable field. What can we do if we can't (or don't want to) use rigorous field searching techniques but still need more control than the fuzzy results free-text searching affords us? Is there any other way we can try to improve the relevancy of our search results?

Proximity commands allow the searcher to describe positional relationships between search terms. They may dictate to the search system the order of ap-pearance of the search words and/or the numbers of words or sentences that may appear between search terms. Although it does not guarantee success in every case, defining the proximity of one term to another is one way to reduce the number of false hits.

Phrase searching is a particular type of *proximity searching* in which two or more words are intended to be searched as a phrase. You would not want the concept "oil spills" to be separated, which could retrieve "kitchen spills can be absorbed using vegetable oil." Systems vary in how they handle phrase search-ing. Some systems assume two words typed without an operator between are to be searched as a phrase, others insert an implied AND (or worse, an implied OR), while others require a proximity operator. A common, but not universal, convention for phrase searching is to enclose the phrase in quotation marks. Proximity commands direct the system to retain the integrity of phrases.

There is usually an on-line help feature that identifies each proximity func-tion allowed in that system. Selected examples of commands from different search systems are listed in figure 6.1. To ensure our search for the exact phrase "oil spills," we might use one of the following search statements:

oil adj spill?
oil pre/1 spill?

A search strategy employing proximity commands including both of our re-search components might look like this:

(otters or enh*dra) same (oil adj spill?)

Figure 6.1. Advanced search commands used in many electronic databases.

Commands	Function	Example	Retrieves
adj	Terms appear next to one another, in exact order typed	spectral adj wavelength	spectral wavelength
near, n	Terms appear next to one another, in any order	inorganic near chemistry	inorganic chemistry chemistry, inorganic
pre/#, adj#	One term precedes the other, by as much as # words	Stephen pre/2 Gould	Stephen Gould Stephen Jay Gould
within#, w/#, n#	Terms appear within as many as # words to each other in any order	pollution within2 oil	oil pollution pollution from oil
same, sames, sent, w/s	Terms appear within the same sentence or field	oil same otters	The otters were covered with oil.
w/p	Terms appear within the same paragraph	oil w/p otters	The oil spill was very detrimental to the local environment. Sea otters were particularly affected by the spill.

This search strategy instructs the system to look for the common or scientific name of the animal and looks for the phrase "oil spills" in the same sentence.

The reasoning behind this design is that if both terms are in the same sentence there is a better chance that the document is relevant to the research topic. The previous example is not the only possible strategy that you could use. Depending on the database search system and how sophisticated you care to be, there are many search strategies available that will provide you with good results. Databases vary widely in the *proximity operators* available and the format required; refer to the on-line help screens to determine what is available in a particular database.

Limiting the Search and
Special Field Searching

Up until now, we have been designing search strategies using subject headings and/or keywords. There are other ways to refine searches as well. Some of the more common ways to *limit* the scope of a search are to restrict it by date, language, type of publication, or geographical area. Many database systems allow one or more of these limits to be placed on the search. By restricting to date range, we can limit our results to the most recent articles for example, the past five years or to an earlier block of years, in order to review the progress of research in a particular field.

Restricting language options should be done cautiously. Although the searcher may not be fluent in more than one language, research progress is nonetheless being made around the world. If you need to be comprehensive in your search efforts, keep in mind that translation services are often available, especially on college and university campuses.

With some variation, science databases such as BIOSIS, Chemical Abstracts Service (CAS), GeoRef, and INSPEC offer a wide selection of document or record types with which to limit one's search. Common types are journal article, book, conference paper or proceeding, and dissertation. In addition, you may be able to further limit to type of article such as review, research, or survey. Other types of documents may include *laws* or *patents.*

Science indexes and databases also developed specialized field searching to augment the usual author, title, subject heading, and key-word searching found in multidisciplinary databases and indexes. For example, the CAS Registry Number is a searchable field in BIOSIS, GeoRef, and others as well as CAS's own database because of the interdisciplinary use and value of the chemical registry numbers. Subject-specific classification codes have been defined in these databases to improve the accuracy of searching over keyword and even authorized subject heading terms. A4710 is the classification code for General Fluid Dynamics Theory in INSPEC; 75202 represents the class of Arthropoda called Chilopoda (centipedes) in BIOSIS; 11 and 12 are the GeoRef codes for Vertebrate Paleontology and Stratigraphy, respectively. In Aquatic Sciences and Fisheries Abstracts (ASFA), Q1 01461 will net records having to do with plankton.

Other fields may allow the searcher to restrict to human or nonhuman experimentation in the biological and medical databases. INSPEC offers a treatment field wherein the researcher may select from theoretical or mathematical, practical, experimental, or application type papers. ASFA provides a field to define the environmental regime such as marine, freshwater, or brackish.

These special fields afford the user an exciting opportunity to prepare a very sophisticated search strategy that will hopefully produce a much more relevant and manageable set of results.

Searching Full-Text Resources

Tight budgetary times are reflected in the purchasing power of the library. Publishers and information producers, who have recognized this problem, have tried to develop products that give the library more "bang for the buck." Many electronic databases now include access to the complete text of the article or document in addition to just citing or indexing it.

The advantages of electronic full-text resources are that you don't have to worry about whether the paper copy of the journal is owned by the library or, if it is owned, whether the library copy is missing or mutilated. Full-text products deliver materials from a wider variety of sources than the library may have been able to afford otherwise. Obtaining your information from a full-text database can be faster than photocopying from print or microform or obtaining a copy through your interlibrary loan service.

While it is useful to have the full-text of an article readily available, searching full-text resources can be problematic. Check to see if the full-text database contains the entire article or just the text of the article. Many databases provide the text of the article but do not include tables, charts, or illustrations. Depending on your subject area and the purpose behind your research, this may or may not be of great concern to you.

Some full-text sources are not well indexed, which makes field searching difficult or impossible. In those cases, free-text searching takes on nightmarish proportions. We saw how free-text searching a well-indexed bibliographic database can produce false hits; imagine the results of free-text searching in a database consisting of millions of words!

The use of proximity commands can greatly improve the efficiency and effectiveness of your full-text search by describing the positional relationship of your search terms. The closer they are in a sentence or paragraph the better the chance that the article you retrieve will be relevant.

Truncation and wildcarding are options too, as long as extra care is taken to think about the ramifications of substitution. Remember, if you truncate a word, it will look for all the variations of that word stem throughout the entire text of the database. Explore what limitations you are allowed in each full-text source you search. These *limiters* may help considerably.

Summary

- Advanced search techniques empower the researcher and offer the flexibility needed to find appropriate materials. Check on-line help for a review of advanced searching techniques allowed in the database.
- Truncation symbols may be substituted for none, one, or several characters at the end of a word.
- Wildcards allow character substitution within a word.

- Proximity commands define the positional relationships between search terms.
- Truncation and wildcard symbols may be defined differently
  within different databases.
- Limits or limiters may be used to refine a search by date, language, or publication type.
- A full-text database provides the entire text of the document in
  addition to a citation or abstract.
- Selection of concise search terms and the use of advanced
  search techniques are critical to successful full-text searching.

Chapter 7

Moving Along: Locating Books

Scientists rely heavily on the timeliness of personal communication, scientific meetings and conferences, and journal articles to keep abreast of the latest advances in their fields. Books, however, afford the student the opportunity to gain knowledge of the history and development of a particular field of scientific study. Without a firm grasp of the knowledge that has already been gained, the student cannot fully appreciate or hope to contribute to the body of new information being gathered through current observations and experimentation.

Although the printed book may not be viewed as the timeliest way to get information, books have not been replaced in the revolution that has been taking place in the information field. Beginning with the wave of computerization in libraries and the publishing industry in the mid-1980s, the book and, especially, the methods for locating books, have changed. For years, the card catalog offered the library user limited access to the library collection. These print-based catalogs have been replaced by electronic catalogs available through personal computers in most academic, public, and special libraries today. The computer offers much more sophisticated and comprehensive access to the library collection. In addition, the book itself is appearing in various electronic formats.

OPACs

To appreciate *On-line Public Access Catalogs*, or *OPACs*, it helps to review the old card catalog system. The three main *access points*, or sections, of the catalog were author, title, and subject headings. Cards within each section were arranged alphabetically, or, in the case of small collections, the author, title, and subject heading cards may have been integrated. Author cards were organized according to the last name of the author or editor of an item. *Location* within the library, as well as other *bibliographic information* about the book, might be available on both the author and title cards. If looking for books by topics, one used the subject heading section. Subject headings used were selected from authorized thesauri such as the *Library of Congress Subject Headings* (see chapter 4). From here, the user might find cross-references to authors or titles pertinent to that topic. Maintenance of the card catalog required considerable physical manpower. The best feature of the card catalog was probably that the user could browse the entries if one wasn't sure of spelling, names, or appropriate headings. In some research libraries older materials that were acquired before the advent of computer-based catalogs may still be indexed in a card catalog. Materials in the card catalog may be important if you are doing historical research on a topic.

A significant advantage to computer-based catalogs is that they expand the points of access. In addition to author, title, and subject heading, other bibliographic information about the book, such as publisher, date, descriptive terms (other than authorized subject headings), language, and call numbers, may be searchable as well. While the OPAC is totally unforgiving when it comes to spelling, it may also permit certain types of browsing. Other features unique to most computer-based OPACs are the ability to provide the circulation status of an item, that is, whether it is available or checked out to another patron, and the ability to *mark* records for printing and emailing. Web-based OPACs allow you to click the mouse on links and commands rather than relying on typing skills and search command languages. Because most academic organizations in the United States have adopted this format, examples used in this chapter concentrate on Web-based OPACs. Of tremendous benefit to students is the ability to freely access many university and college OPACs from around the world.

The way an OPAC displays on the screen of a computer monitor depends on the software used to manage the whole collection. Regardless of the software, many OPACs offer similar search features. Therefore, if you can manipulate one system, it is usually fairly easy to translate that knowledge when searching another system.

The Bibliographic Record

Part of the secret to performing an efficient and effective search is in understanding what you are searching. An on-line catalog (or any library catalog for that matter) is a list comprised of *bibliographic records*, one for each item in a collection. The bibliographic record is the basic unit or level in the on-line catalog. Each record consists of publishing and cataloging data that identifies and describes a particular book or item. These data are organized into separate fields or areas. For example, there is a separate field for author or editor information, the book title, publisher name and location, date of publication, and so forth. Figure 7.1 illustrates the format for a bibliographic record as it would appear in a typical Web-based OPAC.

Figure 7.1. Example of an on-line bibliographic record.

Author, etc.:	Millero, Frank J.
Title:	Chemical Oceanography / Frank J. Millero, Mary L. Sohn
Publisher:	Boca Raton: CRC Press, c1992.
Description:	531 p.: ill. (some col), maps; 25 cm.
Notes:	Includes bibliographical references and index.
ISBN:	0849388406
Subjects:	Chemical oceanography.
Other author(s):	Sohn, Mary L., 1952-

LOCATION:	**CALL NUMBER:**	**STATUS:**
Chemistry library circulating collection	GC111.2 .M55 1991	On loan, due: 5/16/2001

The information contained in any one bibliographic record may vary considerably depending on the item it describes. The ease of interpreting the record may depend on how much information is contained and any abbreviations used. In the first field, the *primary author* is identified. In the title field, the "/" mark separates the full title of the book from the area that is called the "statement of responsibility." In the example in figure 7.1, it is here that we first note that the book has a second author. The place of publication (a state or country may or may not be specifically mentioned), the name of the publisher, and the copyright date of the book are found in the publishing information field. The description provides the number of pages, indicates that there are illustrations (ill.), some of which are in color (some col), and maps. The physical height of the book is 25 centimeters. Notes may be included to further describe the content of the book, in this case pointing out that the book includes additional references and is indexed. You may find the references helpful in locating more information on the topic. A well-indexed book helps you locate specific facts more quickly.

ISBN stands for *International Standard Book Number*, a unique number that is assigned to any published item. The number, in part, represents the publisher responsible for the book and helps differentiate the book from others by the same publisher or books with similar titles. Sometimes students confuse the ISBN number with the call number of a book. A call number, which will be discussed later, indicates the address or location of a book on the shelf in a library collection.

One or more subject headings may be assigned to a book. Just above the location field, we find another note field for additional authors. As you can see, some of the information on the record may be repetitive.

The location field, which appears at the bottom of the record, identifies the library within the university where the book may be located and also indicates where, within the library building, the book can be found. This area also provides the address (call number) for the book, and indicates whether it is currently available for check out. In the example in figure 7.1, the book is part of a circulating collection, which means that it may be borrowed, but it is currently checked out to someone else and provides the date the book is to be returned.

To reiterate, when you search an on-line catalog, it is the bibliographic record with its delineated fields of information that you are actually searching. When a search strategy is entered, the computer retrieves records that match the search parameters. If only one record matches the search, the system will usually display the bibliographic record for that item. If, however, the system retrieves more than one record, it organizes those records into some type of intermediate display, depending on the number of hits and the type of search performed.

Searching the OPAC

The principles related to composing a search strategy, selecting search terms, employing Boolean commands to combine those terms, truncation, and field and free-text searching (chapters 5 and 6) may apply when searching an OPAC. Many on-line catalogs provide both a simple and an advanced search method. The simple search screen may have one box within which to enter a term or terms and a menu that displays a few of the most frequently used search fields. Figure 7.2 illustrates a typical basic search screen in an OPAC.

Figure 7.2. Basic search screen in a typical on-line catalog.

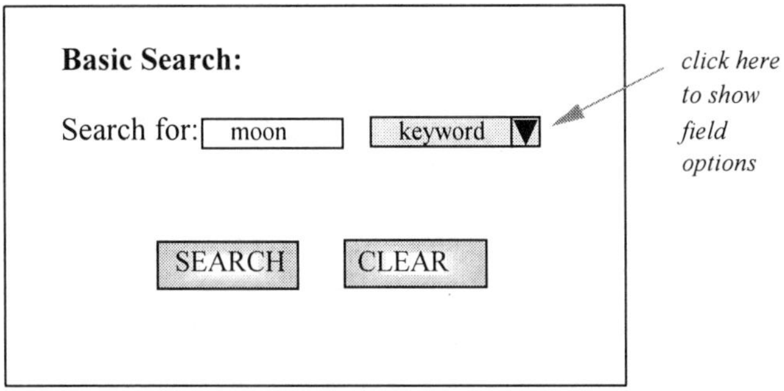

On-line catalogs often employ *pull-down* or *drop-down menus* that allow you to select the specific field that you want to search. You select one of the search types, usually by clicking on the word and highlighting the selected field. You then enter a search term into the box next to the selected field. The system searches the library collection for any record that contains the term or phrase entered in the search box, in the field selected. The search is implemented by clicking on the "Search" box. The basic search screen shown in figure 7.2 has defaulted to a keyword search. This means the computer will look for the word "moon" in *every* area of the bibliographic record.

If you wanted to limit a search to books by an author with the last name of Moon, simply click on the arrow to bring up the menu of field selections. Highlight "author" and submit the search. Figure 7.3 shows an author search for Moon. To search for books using the subject heading "moon," select the Subject field, then type the word "moon" in the search term box and submit the search. Any book that has been assigned the subject heading "moon," will be listed on a results page.

Figure 7.3. Basic search screen illustrating various field searching options.

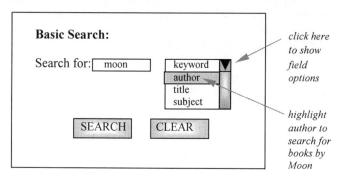

An advanced, expert, or guided search method usually provides more searching options, as exemplified in figure 7.4. Multiple search field and search term boxes may be available for more combinations of field and term searching. There may be more sophisticated field options available as well. Most advanced search screens allow you to limit your searches by date, language, format, and location. The material type or format limit may be particularly useful since this allows you to combine your search term or terms with a limit to things such as video recordings or slides. You may want to limit a search to an *exact match of your typed search entry* such as in the author search:

Author	Williams, Peter

An exact title or subject-heading search would appear as:

Title	Lunar sourcebook a user's guide to the moon

Subject	moon—handbooks manuals etc

You may prefer to limit your search to one field but not be quite as rigorous as the preceding exact search examples. In this case, you might select an author, title, or subject keyword field. In this type of search, the terms must be present in that designated field but in any order. The option to perform an open free-text search may also be available. Additional searchable fields may include note, abstract, call number, or ISBN and other publishing data.

Figure 7.4. Advanced search screen in a typical on-line catalog.

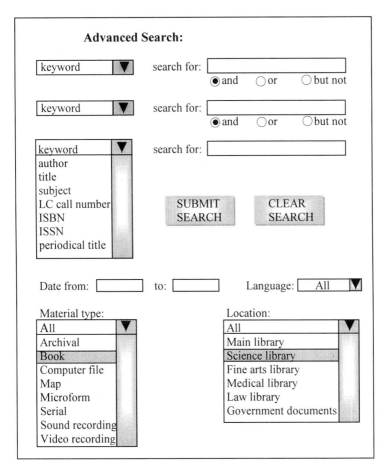

The number of items recovered from a search depends on the size of the collection and, as stated previously, the display of the results depends on the number of hits and the type of search. If the retrieval is small, the results page may display actual titles of books. In performing a subject search, if the retrieval is large, you may see some kind of intermediary result page such as the one shown in figure 7.5. Rather than going directly to a very long list of book titles, the system displays the number of records found under the main subject heading, and subsequently, under any *subheadings* under the general heading. Subheadings are listed alphabetically under the general heading.

In this example, the intermediary results page shows that the subject search resulted in recovery of 194 records, where the term moon was found somewhere in the subject heading field. The numbers on the right indicate how many books fall under the main subject heading and subheading categories. Seventy-four books were assigned the general heading of "moon," three books were assigned "moon—early works to 1800" and so forth.

Figure 7.5. Subject search results in a typical on-line catalog.

Search: subject=moon	Results: 194 titles
Number of Titles:	Subject Heading:
74	MOON
1	—AMATEURS MANUALS
2	—CONGRESSES
3	—EARLY WORKS TO 1800
31	—EXPLORATION
1	—HANDBOOKS MANUALS ETC
5	—MAPS
4	—OBSERVERS MANUALS
2	—ORBIT
1	—ORBIT—MEASUREMENT
2	—ORIGIN
1	—PHASES
3	—PHOTOGRAPHS FROM SPACE
3	—PICTORIAL WORKS
1	—SURFACE

To move forward to a page of specific book titles, you would select one of the subject headings. Web-based OPACS may allow you to click on a highlighted or underlined link to proceed to another page in the catalog. If the OPAC is text-based, you normally enter a line number corresponding to the desired entry. Another intermediary screen may appear, which lists titles and condensed bibliographic citation information for books or documents that are classified under the selected subject heading. This screen may provide enough information to allow you to locate the book or document on the shelf. An example of an intermediary screen listing book titles is shown in figure 7.6. From the title list, the user may then select a specific title.

Figure 7.6. Intermediary search results screen in a typical on-line catalog.

SEARCH: SUBJECT=MOON			
Title:	**Author:**	**Date:**	**Location/Status:**
1. The conquest of space.	Ley, Willy	1949	SCIENCE Lib. TL789.L49 1949 On loan, due 05/15/2001
2. The Earth and its satellite.		1971	MAIN circulating QB631.E28 1971 Not checked out
3. The Earth and its satellite.		1971	SCIENCE Lib. QB631.E28 1971 Recalled, due, 05/01/2001
4. Earth, moon, and planets.	Whipple, Fred Lawrence	1968	MAIN circulating QB601.W45 1968 Not checked out
5. Full moons.	Katzeff, Paul	1980	SCIENCE Lib. GR625.K27 Not checked out

To access the full bibliographic record from the title lists, select the specific record by clicking on the title, clicking on the "full record" *icon*, or entering the line number that corresponds to the desired title. The unique and complete bibliographic record for an individual book is then displayed. Figure 7.7 illustrates the format for a complete bibliographic record. Some OPACs give the user a choice between a shortened bibliographic record including a few of the most important elements of a citation, e.g., the author, title, and call number, or a long record.

Figure 7.7. Full bibliographic record for a book listed in a typical on-line catalog.

Author, etc.:	Ley, Willy, 1906-1969.
Title:	The conquest of space. Paintings by Chesley Bonestell.
Publisher:	New York, Viking Press, 1949.
Description:	160 p. illus., plates (part col.) 28 cm.
Subjects:	Interplanetary voyages.
	Planets.
	Moon.
Other author(s), etc.: a	Bonestell, Chesley, illus.

LOCATION:	**CALL NUMBER:**	**STATUS:**
SCIENCE Lib.	TL789.L49 1949	On loan,
circulating collection		Due 05/15/2001

Additional links may be located within the bibliographic record. To locate similar materials, simply click on one of the subject headings listed, such as "interplanetary voyages," and instantly perform a new search on that subject phrase. It is also possible to perform a quick author search by clicking on the link associated with one of the author's names.

Exact author searches normally follow a standard format. Most OPACs require that you type in the last name of the author first, followed by the first name or first initial if it is available. It is also possible to perform a search on a *corporate author*. If a book or report has been written by a number of people at an organization, the authorship may be attributed to the organization rather than to one or two specific individuals. To locate these materials, use the author search field and enter the organization name such as "university of texas at austin bureau of economic geology" or "american chemical society."

An author search usually results in a display of author's books owned by the library, although in the case of a very prolific author, the intermediary display may be author headings from which you move to a title list. To illustrate this, try searching for the author Isaac Asimov on the OPAC of a large science library.

Don't forget that you can also perform subject searches on a particular person. If you want to locate books *about* Charles Darwin, you would use a subject search of "darwin charles." To find materials that were written *by* Charles Darwin, you would use an author search.

An exact title search generally results in a single bibliographic record display. An initial "a," "an," or "the" in a title is not considered part of the title, but these words must be included in a title search if they appear after the first word of the title. To search for a book titled *The Crust of the Earth*, you would enter an exact title search of: "crust of the earth." Note that the first "the" is deleted from the search, but the second "the" is included in the search statement. Remember, OPACs are unforgiving about incorrect entries! A keyword title search will generate a list of books that contain the search term somewhere in the title.

In addition to the display of search information, most on-line catalogs provide a menu or toolbar of commands somewhere on the display screen, which enable you to move throughout the various levels of the on-line catalog and perform functions previously unavailable with a print catalog. Figure 7.8 illustrates a toolbar with some of the commands and options that may be available on your OPAC.

Commands in these menus are often screen specific; in other words, they may change depending on what function you have just performed. Options appearing on an initial search screen, a results list page, or a bibliographic record screen may vary since some commands are only relevant at specific stages of a search. Commands such as "search results" will move you from an individual bibliographic record back to an intermediary list of titles. "Last Search" allows the user to edit the last strategy entered rather than starting all over again. Some OPACs have a search history option that allows you to view all previous searches performed and recombine search steps to create new searches.

Marking records to print helps to ensure that the item will be correctly cited and to facilitate locating the materials, as well as providing a tracking mechanism for sources used. E-mail capabilities allow you to download to bibliographic management software such as *Procite* or

EndNote for bibliography maintenance. It also assists in the sharing of information between students, professors, and other colleagues.

Figure 7.8. Toolbar listing options that may be available in a typical on-line catalog.

Search
Options
_____

Help
Search Results
Last Search
Search History
New Search
Request on ILL
Mark
E-mail Marked
Print Marked
Unmark

As mentioned in the introduction to this chapter, on-line catalogs of many universities and colleges are openly available through the Internet. The ability to mark catalog records and print citations or to e-mail them has greatly facilitated the acquisition or borrowing of materials. Many students can do their preliminary searching for materials from a place other than the library, such as classrooms, laboratories, residential halls, or home.

Book Databases

It is impossible for any one library to acquire every book published. If your library does not have enough books on a particular topic, there are ways to discover additional sources. As mentioned previously, students now have access to the catalogs of other educational and institutional organizations. There are also commercially prepared products that may be helpful. A major directory of materials published in North America or distributed by U.S. companies is called *Books in Print (BIP)*. Available both in print and electronically, *BIP* provides publishing information for both recently out-of-print and currently in-print materials. The directory is searchable by several fields, including subject, title, author, and ISBN.

Another helpful commercial product is *WorldCat*, a catalog of millions of items including books, *serials*, government documents, maps, media, and other materials amassed by *Online Computer Library Center, Inc. (OCLC)* member libraries worldwide including most academic and research libraries. Both *Books in Print* and *WorldCat* are easy to search. Once the relevant books have been located, you may check with your interlibrary loan office to find out if the materials are available for loan.

For years, individual publishers have produced their own print catalogs for distribution to libraries and booksellers. Many of these catalogs are now available to the general public through company Web sites, such as the popular Amazon.com, http://www.amazon.com and Barnes and Noble, http://www.bn.com.

Several well-known scientific publishers offer access to their book, periodical, and other media catalogs free of charge. Some of these publishers include Blackwell Science, http://www.blackwellscience.com, John Wiley & Sons/Van Nostrand Reinhold, http://catalog.wiley.com, and McGraw-Hill, http://www.mhhe.com/catalogs/sem. National Academy Press also provides open access to its collection of reports and publications from the National Academy of Sciences, the National Research Council, and the National Academy of Engineering, to name just a few. You may search these catalogs by author, title, subject headings, or topical areas. The catalogs provide complete publishing information and availability, a description of the book, and, in many cases, a table of contents. Publishers' catalogs also advertise new and forthcoming books.

Dissertations and Theses

Dissertations and theses are specialized books. The university where a graduate student completed his or her master's or PhD research publishes these materials, and they may be difficult to obtain. In most cases, access to dissertations and theses is not necessary for an undergraduate science paper since these materials tend to be very focused. Students who are planning to attend graduate school must be aware of these resources, however, since it is important to know what research has already been accomplished prior to selecting a topic for your own graduate research.

Dissertations are most easily located by searching a database called *Dissertation Abstracts*. This database includes author, title, and keyword access to PhD dissertations that have been published in North America and some European countries since 1861. Indexing to some master's theses is also included. The database is somewhat difficult to search by subject since it does not have a very refined controlled vocabulary. Subject headings that are assigned are extremely broad. The database can be searched by keyword, but you must try to consider every possible form or synonym that may be related to your search topic. Dissertations and theses may also be cataloged and included in the *WorldCat* database or they may be indexed within subject-specialized databases such as *GeoRef* or *INSPEC*.

Dissertations and theses may be difficult to obtain since there are a limited number of copies available for each title. Many dissertations and theses are available for purchase from University Microfilms, Inc. (UMI). It is also possible to check with the library at the institution that granted the graduate degree. Many academic libraries will lend a dissertation or thesis on interlibrary loan if they have a second circulating copy.

E-books

While the electronic availability of journals has been marketed successfully for several years, electronic books, or *e-books*, are more recent. Designers continue to experiment to find the most appropriate technologies to deliver a book electronically. Other issues that have been hotly debated deal with ownership, copyright, length of time accessible, quality of product, and market demand. As with any new idea, there have been a variety of resulting products and services that have appeared in the publishing and library market. Emerging as potentially attractive resources to the science student are e-book collections.

There are open-access collections of electronic books such as the *On-line Books Page*, http://digital.library.upenn.edu/books/ and the *Internet Public Library (IPL) On-line Texts Collection*, http://www.ipl.org/reading/books. These collections include many materials produced by government agencies such as the National Research Council and the United States Geological Survey (USGS). Other documents are accessible from university sites or are older books and materials that are no longer bound by copyright laws. Among the latter category, Project Gutenberg, http://promo.net/pg/ contains many important early scientific works.

Fee-based commercial e-books are available by subscription to libraries or individuals. The science content of a major e-book collection called *netLibrary* includes books from scientific publishers such as Blackwell Science, Cambridge International Science, Marcel Dekker, John Wiley, Kluwer Academic, McGraw-Hill, Oxford University Press, the Smithsonian, and many university presses.

Using the E-book Collection

In the traditional library practice, the student finds a book on the shelf, pulls it down, and takes it to a circulation desk where it is checked out for a certain period of time. The book is returned (or should be anyway) by or before the due date.

An e-book database may allow you to search for books by author, title, keyword, date, ISBN, and publisher, much as you would in a library OPAC. Once an e-book has been selected, the table of contents appears, as well as the text of the book itself, including any introductions, acknowledgments, indices, glossary, bibliography, and appendices. You may begin to browse the book or to perform another search to look for specific content in much the same way you would use the index of a print book to locate the pages where a particular topic is discussed.

You may be required to establish a user profile in order to read the entire content of a book on-line or to take advantage of the option to download a file copy of the e-book to a local computer. The software needed to open an e-book file is usually available from the e-book provider and may be maintained on a local computer for future borrowing. Downloading the book may provide the reader with additional options to highlight areas for note taking or mark specific sections of the book for printing. The reader may also make *annotations* to sections of the text, and save both the text and the notes to a separate file. These annotations are not seen by other borrowers, but are available to you if you check out the same book again.

There may be a time limit imposed on a borrowed e-book just as there is for the print library book. The time allowed is determined by the organization that contracts with the e-book provider. After the time limit is up, a downloaded file becomes unreadable and can be deleted from the local computer.

Classification Systems

Most scientists are familiar with classification systems of one kind or another. Biologists classify living organisms, chemists classify chemical compounds, and geologists love to organize rocks and minerals. Library materials and information need to be organized as well if the patrons of the library are to be able to find and utilize the library's collection.

There are several different classification systems available for organizing library collections. School children and public library users are most familiar with the *Dewey Decimal Classification System.* These systems provide a method to design a number that will be unique to every book within a collection. This number is known as its *call number.* You use the call number to locate the item on the shelf, as you would use a street address to locate a house within a city.

Figure 7.9. Dewey Decimal classification categories for science and technology.

500		**Pure Sciences**
	510	Mathematics
	520	Astronomy
	530	Physics
	540	Chemistry
	550	Earth Science
	560	Paleontology
	570	Life Sciences
	580	Botany
	590	Zoology
600		**Applied Sciences & Technology**
	610	Medicine
	620	Engineering
	630	Agriculture
	640	Home Economics
	650	Management
	660	Chemical Technologies
	670	Manufacturing
	680	Application-specific Manufacturing

The call number may identify either the subject area of the book, the author, the title, or some combination of these. In figure 7.9, the Dewey Decimal classification numbers are listed for the areas of science and technology. While it is not our intent to convert science students into librarians by detailing the ins and outs of library cataloging, it may be very helpful for any student to be able to recognize the general call number area for subjects in which he or she may concentrate.

Many academic and very large public library systems use the *Library of Congress Classification System* (*LC*). The system was developed to offer more options and flexibility that a larger and more complex library collection may require, while using a shorter call number. In LC, one or two letters represent broad subject areas as illustrated in figure 7.10. To better understand the composition of a call number, let's walk through the steps for classifying a book. Consider the following work:

Kenyon, Karl W. *The sea otter in the eastern Pacific Ocean.*
New York: Dover Publications, 1975.

Since sea otters are animals, they fall under the subject area of zoology represented by the Library of Congress classification QL. A number from 1 to 999 further defines zoology into types of animals. So, for example, the range of QL 700-739.8 refers to mammals. Within this range, a specific number-letter combination represents different orders and families of mammals. For example, the otter is a family called *Mustelidae* within the order of carnivores, and is assigned the classification number QL 737.C25. This completes the *class stem* of the call number.

From here, a specific item within the class is designated with a *book number*. The author, title, or date of a book may be represented in the number, starting with the addition of another letter/number combination. The author (Kenyon) of the book may be represented by adding the first letter of the last name, in our case K, to the call number, followed by one or more numbers to stand for subsequent letters of the name. While it may seem arbitrary, there is actually a table that library catalogers use to generate this combination and assure some measure of consistency among libraries. Adding the publication date to the call number differentiates it from other editions. The final result is a unique, Library of Congress call number for the book in hand: QL737.C25 K4 1975.

Figure 7.10. Library of Congress classification categories relating to science and technology.

G	**Geography (General)**	**Atlases**	**Maps**
	GA	Mathematical Geography	
	GB	Physical Geography (including hydrology)	
	GC	Oceanography	
	GE	Environmental Sciences	
Q	**Science**		
	QA	Mathematics	
	QB	Astronomy	
	QC	Physics	
	QD	Chemistry	
	QE	Geology	
	QH	Natural History - Biology	
	QK	Botany	
	QL	Zoology	
	QM	Human Anatomy	
	QP	Physiology	
	QR	Bacteriology/Microbiology	
R	**Medicine**		
S	**Agriculture**		
	SB	Plant culture	
	SD	Forestry	
	SF	Animal culture	
	SH	Aquaculture, Fisheries, Angling	
T	**Technology**		
	TA	Engineering - General Civil Engineering	
	TC	Hydraulic Engineering, Ocean Engineering	
	TE	Highway Engineering, Roads & Pavements	
	TF	Railroad Engineering	
	TG	Bridge Engineering	
	TH	Building Construction	
	TJ	Mechanical Engineering	
	TK	Electrical Engineering	
	TL	Aeronautics, Astronautics, Motor Vehicles	
	TP	Chemical Technology	
	TR	Photography	

The *National Library of Medicine Classification* is a specialized system for the field of medicine and related areas. The system finds its roots in the *Library of Congress Classification,* but utilizes certain letter combinations that have been excluded from LC (QS-QZ and W-WZ) and eliminates some other LC categories to avoid redundancy.

Figure 7.11. Selected National Library of Medicine classification categories relating to science and health.

	Preclinical Sciences
QS	Human Anatomy
QT	Physiology
QU	Biochemistry
QV	Pharmacology
QW	Microbiology and Immunology
QX	Parasitology
QY	Clinical Pathology
QZ	Pathology
	Medicine and Related Subjects
W	Health Professions
WA	Public Health
WB	Practice of Medicine
WC	Communicable Diseases
WD	100 Nutrition Disorders
WD	200 Metabolic Diseases
WD	300 Immunologic and Collagen Diseases. Hypersensitivity
WD	400 Animal Poisons
WD	500 Plant Poisons
WD	600 Diseases and Injuries Caused by Physical Agents
WD	700 Aviation and Space Medicine
WE	Musculoskeletal System
WF	Respiratory System
WG	Cardiovascular System
WH	Hemic and Lymphatic Systems
WI	Digestive System
WJ	Urogenital System
WK	Endocrine System
WL	Nervous System
WM	Psychiatry
WN	Radiology. Diagnostic Imaging
WO	Surgery
WP	Gynecology
WQ	Obstetrics
WR	Dermatology
WS	Pediatrics
WT	Geriatrics. Chronic Disease
WU	Dentistry. Oral Surgery
WV	Otolaryngology
WW	Ophthalmology

Selected National Library of Medicine classification categories are displayed in figure 7.11.

United States government documents are frequently classified using the *Superintendent of Documents Classification System*, which will be discussed in more detail in chapter 10. While the other classification systems presented here group materials primarily by subject, the government document system organizes materials by the U.S. government department or agency that generates the item.

Summary

The bibliographic record is the basic unit of the on-line library catalog. The bibliographic record consists of publishing and cataloging data that identifies and describes the book.

On-line catalogs (OPACs) may vary slightly from one library to another, but most provide similar options for searching. Basic search screens normally permit the user to search by keyword, author, title, or subject. Advanced, guided, or expert search screen options may allow additional search fields such as the ISBN or call number. An advanced search screen may also provide the opportunity to limit a search by date, language, or format.

A classification system groups books about the same topic together in a library collection, which allows the materials to be located quickly and easily. Several classification systems are available in the United States including the Library of Congress, the National Library of Medicine, the Dewey Decimal, and the Superintendent of Documents systems.

In addition to your own library catalog, you may also search the catalogs of other libraries. *Books in Print, WorldCat,* and publishers' on-line catalogs may also be used to find additional resources that may be available locally. If materials are not available at a local library, check with your interlibrary loan department to find out if the resource may be available for loan.

Electronic books are becoming increasingly popular and often have the advantage of allowing the user to search the entire text of the resource.

Chapter 8

Locating Information in Journals, Conference Proceedings, and Newspapers

One of the most frequently asked questions at the library reference desk is "How do I find magazine articles on my topic?" An article may provide a concise overview of, or an introduction to, a topic—a big time-saver if you don't have to read an entire book! Many other articles focus on a narrow aspect of a subject with greater depth or detail. Most important, articles normally appear in print a *lot faster* than books! You don't have to wait a year or two to read about something that is happening now. As you know, this is critically important in science.

Before we get into the specifics of indexing, let's discuss a little terminology. Many people use the terms *periodical, magazine,* and *journal* interchangeably. A periodical is actually a generic term for magazines and journals. A magazine is a popular type of publication that is intended for the general public. Think of a magazine as the kind of publication that you can easily purchase at a newsstand. They are widely available, and usually contain lots of illustrations and advertisements. Journals, on the other hand, are usually considered to be more scholarly with few, if any, advertisements. Journals normally contain lengthier articles that are more research oriented and nearly always identify the author of the article. Journal articles also often contain bibliographic references and may be peer-reviewed.

The most scholarly type of periodical is the *refereed* journal, which may also be called a *peer-reviewed* or *juried* journal. Before an article is published in a refereed journal, it must pass a peer review process. The article manuscript is sent out to experts in the field of study. If these experts feel that the article has merit, they will recommend it for publication. An even stricter form of peer review is a process called *blind review.* In this case, the reviewer does not know the author of the article. This helps to avoid bias on the part of the reviewer.

Ulrich's Periodicals Directory provides a quick method for determining which journal titles are refereed. A similar reference book, *Magazines for Libraries* by Bill and Linda Katz, also indicates whether a periodical is refereed.

Thousands of periodicals are published every year, so locating an article on a particular topic can seem pretty intimidating. Many folks try browsing the periodical stacks hoping to chance upon an article on their topic. Not a good move! First, it can be *extremely* time-consuming. Most journal publishers do not provide an index for each individual title, so you may end up looking through a large number of individual issues before finding a relevant article. Some libraries shelve their periodicals alphabetically by title rather than by subject, which makes it difficult to determine which journals to browse. Finally, browsing the shelves does not allow you to examine journal titles that are not available in your library.

So if browsing the shelf is not useful, how do we locate periodical articles?

Periodical Indexes and Abstracting Services

The most efficient way to locate articles is through the use of a periodical index or *abstracting service.* Some indexes are general or multidisciplinary in their approach; others concentrate on a general field of study such as business, education, psychology, or science. Many indexes specialize in a specific discipline within a subject area; for instance, science indexes may be focused on biology, chemistry, ecology, computers, or microbiology.

Traditionally, an index provides a list of article *citations* arranged under subject headings. Abstracting services provide citations to the articles as well, but also include a brief summary or *abstract* that describes the content of the article. There are also publications titled "indexes," which include summaries. A little confusing? Well, it can be, so to make our lives a little easier for now, we're going to refer to all of them as "indexes."

All periodical indexes will provide a bibliographic citation for each item listed. A bibliographic citation includes the information that will be needed to locate the material. For a journal article, this includes the author or authors of the article, the title of the article, the title of the journal, the volume and/or issue number for the journal, the complete date of the journal issue, and the inclusive page numbers of the article. In some indexes, only the beginning page number for the article is included. The title of the article is occasionally omitted in the citation, particularly for chemistry journals. While all periodical indexes should include at least this minimum amount of information, they are not necessarily consistent in how they present the information. The introductory material to the index (or the help screens for a computerized index) will show you exactly how to read an entry.

How Are Indexes Created?

The publisher first decides what field (or fields) of study will be included. Does it want to index periodicals in chemistry, in engineering, or in several scientific disciplines? The specific journals that are published in that field or in related fields are then examined to determine whether they will be included in the index. The publisher also decides whether to index the titles comprehensively or selectively; that is, will every article from the journal be cited or will only a few relevant articles from each issue be added to the index? Most indexes will cover the core journals in the discipline comprehensively and additional material that relates to the field of interest selectively from other journals.

Now the indexers begin their work. They read the material, looking for key terms that most effectively describe the contents of the article. Based on subject content, the indexer assigns one or more subject headings to the article. For additional information on this process, refer back to the section on controlled vocabulary in chapter 4.

When the indexers are done, they send the bibliographic citations, the subject headings, and, in some cases, a summary of the article off to be printed or added to the electronic database. Depending on the publisher, the entire process can take a month or more to complete.

Selecting the Appropriate Index

As stated in chapter 4, any time you need help with selecting a subject-appropriate resource, you should ask the reference librarian for suggestions. You may also check the library's catalog to find out what indexes are available in a particular subject area. Search the on-line catalog for your broad subject, and then add the "indexes" subdivision. For example, if you want to locate periodical indexes in the field of chemistry, look up the subject: chemistry—indexes.

A useful resource for locating periodical indexes is a well-known reference tool titled *Ulrich's International Periodicals Directory ("Ulrich's")*. This comprehensive resource is available at many libraries and provides a wealth of information. Each entry gives the name of the periodical; its frequency of publication; subscription information such as price, publisher's address, and telephone number; and a listing of the indexes that cover that periodical title.

The entry in figure 8.1 is for a science journal titled *The American Naturalist*. *Ulrich's* provides all kinds of information about the journal. In this case, you can learn that *The American Naturalist* began publication in 1867, and it is currently issued on a monthly basis. The *Ulrich's* record also gives the subscription price and the publisher's address, telephone number, and e-mail and Web addresses. In addition, *Ulrich's* indicates whether or not the journal is a peer-reviewed (refereed) publication. As we can see from the bottom of figure 8.1, *The American Naturalist* is a peer-reviewed journal, which indicates a high level of scholarship.

Perhaps the most useful feature of *Ulrich's* is the list of indexes that cover this particular journal. Back at the beginning of the chapter, we told you that periodical index publishers select specific titles to cover in their publications. Well, *Ulrich's* is the place to find out which publishers chose *The American Naturalist* for their indexes. Suppose your professor told you that a particularly interesting article on bird nest parasitism had appeared in *The American Naturalist,* but he didn't know the author or the date when it was published. You *could* go to the periodical stacks and start leafing through each issue of *The American Naturalist*. Being the good library researcher that you are, however, you instead proceed straight to *Ulrich's* to look up *The American Naturalist*. You find that this journal is indexed in several well-known periodical indexes including *Biological Abstracts* and *Zoological Record*. You then head to one of these indexes, perform a keyword search on "nest parasitism," and quickly locate the exact citation for the article!

The periodicals listed in *Ulrich's* are organized by broad subject area. *Ulrich's* can also be used for suggestions on indexes or abstracting services that are available in a particular subject discipline. For example, if you are interested in finding out what indexes cover the field

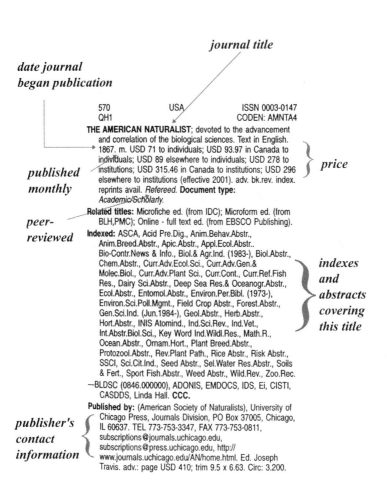

journal title

date journal began publication

published monthly

peer-reviewed

publisher's contact information

570   USA   ISSN 0003-0147
QH1   CODEN: AMNTA4
THE AMERICAN NATURALIST; devoted to the advancement and correlation of the biological sciences. Text in English. 1867. m. USD 71 to individuals; USD 93.97 in Canada to individuals; USD 89 elsewhere to individuals; USD 278 to institutions; USD 315.46 in Canada to institutions; USD 296 elsewhere to institutions (effective 2001). adv. bk.rev. index. reprints avail. *Refereed.* **Document type:** *Academic/Scholarly.*

Related titles: Microfiche ed. (from IDC); Microform ed. (from BLH,PMC); Online - full text ed. (from EBSCO Publishing).

Indexed: ASCA, Acid Pre.Dig., Anim.Behav.Abstr., Anim.Breed.Abstr., Apic.Abstr., Appl.Ecol.Abstr.. Bio-Contr.News & Info., Biol.& Agr.Ind. (1983-), Biol.Abstr., Chem.Abstr., Curr.Adv.Ecol.Sci., Curr.Adv.Gen.& Molec.Biol., Curr.Adv.Plant Sci., Curr.Cont., Curr.Ref.Fish Res., Dairy Sci.Abstr., Deep Sea Res.& Oceanogr.Abstr., Ecol.Abstr., Entomol.Abstr., Environ.Per.Bibl. (1973-), Environ.Sci.Poll.Mgmt., Field Crop Abstr., Forest.Abstr., Gen.Sci.Ind. (Jun.1984-), Geol.Abstr., Herb.Abstr., Hort.Abstr., INIS Atomind., Ind.Sci.Rev., Ind.Vet., Int.Abstr.Biol.Sci., Key Word Ind.Wildl.Res., Math.R., Ocean.Abstr., Ornam.Hort., Plant Breed.Abstr., Protozool.Abstr., Rev.Plant Path., Rice Abstr., Risk Abstr., SSCI, Sci.Cit.Ind., Seed Abstr., Sel.Water Res.Abstr., Soils & Fert., Sport Fish.Abstr., Weed Abstr., Wild.Rev., Zoo.Rec.
—BLDSC (0846.000000), ADONIS, EMDOCS, IDS, Ei, CISTI, CASDDS, Linda Hall. **CCC.**

Published by: (American Society of Naturalists), University of Chicago Press, Journals Division, PO Box 37005, Chicago, IL 60637. TEL 773-753-3347, FAX 773-753-0811, subscriptions@journals.uchicago.edu, subscriptions@press.uchicago.edu, http:// www.journals.uchicago.edu/AN/home.html. Ed. Joseph Travis. adv.: page USD 410; trim 9.5 x 6.63. Circ: 3.200.

price

indexes and abstracts covering this title

Figure 8.1. Main entry for the journal *The American Naturalist* as listed in *Ulrich's Periodicals Directory*, 2001, p. 623.
Reprinted with the permission of R.R. Bowker LLC. Copyright 2000.

of biology, you can turn to the section labeled "Biology—Abstracting, Bibliographies, Statistics." From there, you can go to your library catalog to find out if your library owns any of the titles listed there.

If your topic includes more than one concept, be sure to consider all the subject indexes that may cover the topic. For instance, if you want to find articles about chemical spills in industrial plants, you might use *Chemical Abstracts* for its emphasis on chemistry, *Environmental Sciences and Pollution Management* to find citations to articles on the environmental aspects of the issue, and a business index such as *ABI Inform* to provide a business slant to your research. Don't forget the broader "undergraduate" databases that cover the core journals in all fields; these can be particularly useful for interdisciplinary research problems for which it can be difficult to find precise keywords. It is always helpful to keep your options open.

In printed indexes, an alphabetic listing of the indexed journal titles often appears in the front of the volume or as a supplement to the index. If you want to get a feel for what kind of journals are covered in a particular index, look at the journal title list. In an electronic database, the publisher may include a list in the "about the database" section. If not, check with the reference librarian. The librarian may have a printed list available that corresponds to the electronic database.

Index Features

Armed with the knowledge you have gained from your reference librarian, the on-line catalog, and *Ulrich's,* you have now selected a few indexes that might be useful for your work. Now what? We know that indexes vary in their subject coverage, but they can also vary in a number of additional ways. As we have already discussed, before using any resource, check the introductory material or the help screens to locate a description of the resource, the time period that it covers, and information on the format of the entries.

An important aspect to consider when selecting a periodical index is a determination of the intended audience. Are the journals included in the index scholarly or are they more oriented toward the general public? Depending on your specific needs, this distinction can help to guide your selection.

How good is your Russian? Some indexes are international in scope, covering materials that are published in many countries around the world. Although the article citation and summary are usually written in English, the article itself may be written in Russian, German, French, or any of a number of languages.

The format of the bibliographic citation may also vary from one resource to another. Some include the bare bones, while others add a summary (or abstract) of the article content. Article titles can be misleading so a summary may help to determine if an article is relevant to your topic. Remember, the abstract is included as a guide only. It is not a replacement for obtaining the full-text of the article since it is only a very small portion of the full content and, in some cases, the abstract may have been written by someone other than the original author of the article.

Some electronic databases link to the actual text of the article. The electronic full-text article may be an exact replica of the original printed article complete with tables, charts, graphs, and photos, or it may include only the text of the article. If the graphical images are relevant to your research topic and are not included in the on-line text, you will have to obtain a copy of the original article, either from the periodical stacks or through interlibrary loan.

Using Periodical Indexes

Remember a few chapters ago when we plotted and planned our search strategy? Well, now it is time to use it! It is always useful to keep in mind the subject content of the index you will be using. Let's go back to the search about the chemical spills in industry. The main topics here are:

> chemical spills   AND   industry

related terms could be:

> solvent spills   AND   business

If you use the periodical index, *ABI Inform*, to perform your search, you will want to avoid using the terms "industry" and "business." Everything in *ABI Inform* is related to business, so that concept is automatically covered! You only have to search the terms relating to chemical spills.

Printed Indexes

At this point we should make some distinctions between printed indexes and printed abstracts because their physical format can be quite different. These differences are less significant in electronic versions of indexes and abstracts.

Most printed periodical indexes, such as the printed version of *Applied Science and Technology Index*, are arranged alphabetically by subject. To help to narrow a search, many of the subject headings may be broken down further with subdivisions. As you can see in figure 8.2, the subject "Chemical Spills" is further divided by subdivisions, such as "Cleanup," "Computer simulation," and "Environmental aspects."

You may also notice additional subject headings listed under the *see* or *see also* instruction. In figure 8.2, the subject "Chemical spills" has a see also reference to "Rhine River chemical spills, 1986." If you go to that subject in the same volume of the index, you will locate an additional article or articles relating to chemical spills. These related topics could help to expand your search.

Under each subject heading and subdivision, you will find one or more bibliographic citations. Look at a citation in figure 8.2. In these citations, the title of the article appears first, followed by the name of the author or authors. An abbreviated periodical title is in italics, followed by the volume, pages, and the date of the periodical issue. If the periodical title is abbreviated, be sure to look in the front of the volume to find out the complete name—you can't find out if your library subscribes to the journal if you don't have the complete title. Additional information may also be included in the citation such as "il" (abbreviation for illustrated) or "por" (portrait). The meaning of each abbreviation should be defined in the introductory material for the index. Most printed indexes cite their information in a similar manner. If you are unsure of how to read a citation, be sure to check the front of the index for an explanation of the format used by that particular publisher.

Printed Abstracting Services

Using a printed abstracting service often involves a few extra steps. Remember that abstracting services include a summary of the article in addition to the citation; therefore, each entry may be considerably

main subject heading

reference to additional subject headings

Chemical spills
See also
Rhine River chemical spills, 1986
Comparison of spill frequencies and amounts at vapor recovery and conventional service stations in California. J. J. Morgester and others. *J Air Waste Manage Assoc* 42:284-9 Mr '92
EMERG: an expert system to develop emergency responses to hazardous material spills. W.-Y. Lin and P. Biswas. bibl flow charts diag *Hazard Waste Hazard Mater* 8:263-74 Summ '91
Hazardous materials. See issues of Fire Engineering
Hazardous materials in transit: a public health concern. A. Knill. bibl *Prof Saf* 36:40-2 N '91
Initiation of fire growth on fuel-soaked ground. H. Ishida. bibl (p228-30) il diags *Fire Saf J* 18 no3:213-30 '92
Rail benzene spill forces major evacuation. M. Reisch. *Chem Eng News* 70:5-6 Jl 6 '92
Remote sensing zeros in on river spill. *Civ Eng (Am Soc Civ Eng)* 62:20 Ag '92
Truck hits bridge and derails train. il *ENR* 227:9 D 2 '91
Wisconsin train accident costly, but not deadly. M. Reisch. *Chem Eng News* 70:18 Ag 3 '92; Discussion. 70:3 Ag 24 '92

subdivision of the main subject

Cleanup
Do's and dont's for cleaning spills. R. W. Duncan. *Pollut Eng* 23:72 N '91
Health and safety for PCB remediation. K. E. Fischer. il diag *Pollut Eng* 24:76-8 F 1 '92
Neutralizing a sulfuric acid spill. J. E. O'Neill. il maps *Water Environ Technol* 4:56-61 Jl '92
Remediation of leaking USTs: a system for accessing case histories and related documents. R. W. Hillger and R. A. Griffiths. bibl *J Air Waste Manage Assoc* 42:298-302 Mr '92
Removal of chemical contamination from vehicles: a comparison of weathering and active clean-up processes. D. Amos and B. Leake. bibl *J Hazard Mater* 32:105-12 S '92

Computer simulation
Management training software system simulates disaster situations [CriSys] D. O'Sullivan. *Chem Eng News* 70:21-3 S 21 '92

Environmental aspects
Modelling heavy gas cloud transport in sloping terrain. J. Kukkonen and J. Nikmo. bibl (p175-6) diag *J Hazard Mater* 31:55-76 Jl '92
Recovery of heterotrophic soil bacterial guilds from transient gasoline pollution. M. S. Diltz and others. bibl (p272-3) *Hazard Waste Hazard Mater* 9:267-73 Fall '92

subdivisions of the main subject

Figure 8.2. Excerpt from the paper version of *Applied Science and Technology Index*, 1992, p. 349.

longer than a simple bibliographic citation. Citations are normally *not* listed directly under the subject headings in the *index section* of the abstracting service. Rather, you are given an *accession number* (numbers or letter-number combination) and, in some cases, a brief citation or description of the article. The accession number is just a unique number assigned to each article indexed. Don't let it throw you.

Figure 8.3 illustrates a page from the subject index section of an abstracting service titled *Oceanic Abstracts (Cambridge Scientific Abstracts)*. Subject headings are in boldface type, followed by an alphabetical listing of article titles and a number. If we look at "sea grass," we find an entry for an article about the sea grass environment in Tampa Bay, Florida. The article's number (the accession number) is 3672. This is the number that we use to locate the *main entry* for that item in the abstract section of *Oceanic Abstracts*.

Take the accession number and go to the main entry section (this may be called the *abstract* section). The main entry for each record contains the accession number, the bibliographic citation information, and the summary of the article. Depending on the abstracting service, this may be in a different part of the same volume or it may be in a completely separate volume. If the index and main entry sections are separate volumes, be sure to select the correct date or volume when moving from one section to another.

Using the accession number 3672 that we located under "sea grass" in the index, we find the main entry for this article, as illustrated in figure 8.4. The abstract section is arranged numerically by the accession number. In some abstracting services, such as we see here in *Oceanic Abstracts*, the accession number has some additional numbers attached to it. Although not listed that way in the index, the accession number appears in the abstract section as: 99-3672O. Not to worry, the "99" simply refers to the publication date of the index and the "0" at the end of the number tells us that the citation came from *Oceanic Abstracts*.

Examine the main entry for the sea grass article. The citation information is given first: article title, authors, the authors' affiliations, title of the journal, volume and issue numbers, date, and pages. In this resource, the title of the journal is provided in full and there is no need to refer to a separate list for that name. The paragraph that follows is the abstract or summary of the article. Remember, the abstract is intended as a guide only. We may not know who prepared the summary,

Scientific personnel,

Deep-sea biodiversity: a tribute to Robert R. Hessler 3649

Dr N K Panikkar: learning research by doing 3742

Management of marine natural resources through by biodiversity
informatics 3764

Scouring,

Effects of a high rise building on wind flow and beach characteristics
at Atlantic City, NJ, USA 4391

Sea grass,

Artificial shelters for spiny lobster *Panulirus argus* (Latreille): an evaluation of occupancy 3441

Inhibitors and activators of endo-1 → 3-β-D-glucanases in marine
macrophytes 3357

Response of shoal grass, *Halodule wrightii*, to extreme winter conditions in the Lower Laguna 3690

Seasonal and age-dependent variability of *Posidonia oceanica* (L.)
Delile photosynthetic 3360

Settlement and recruitment of queen conch, *Strombus gigas*, in seagrass meadows: Associations 3411

Spatial and temporal variation of marine bacterioplankton in Florida
Bay, U.S.A. 3693

Studies on germination and root development in the surfgrass *Phyl-
lospadix torreyi*: implications 3353

Sub-lethal effects of coastal petroleum pollution on *Spartina alterni-
flora* 4193

The influence of habitat structure in faunal-habitat associations in a
Tampa Bay seagrass system 3672

subject heading

accession number

Figure 8.3. Excerpt from the subject index of *Oceanic Abstracts*, April 1999, p. 223.

accession number

99-36720 The influence of habitat structure in faunal-habitat associations in a Tampa Bay seagrass system, Florida. Knowles, L.L.; Bell, S.S. (Department of Ecology and Evolution, State University of New York at Stony Brook, Stony Brook, NY 11794-5245, USA) *BULLETIN OF MARINE SCIENCE,* Vol. 62, No. 3, May 1998, pp. 781-794, ISSN 0007-4977 *Published by:* ROSENSTIEL SCHOOL OF MARINE AND ATMOSPHERIC SCIENCE. En;en.

citation

This study examined the distribution patterns of epifaunal associates of two species of seagrass, *Syringodium filiforme* and *Ruppia maritima*, and two drift algae, *Gracilaria* sp. and *Spyridia* sp. Pericarid crustaceans (i.e., amphipods, isopods, and tanaids) dominated all macrophyte collections. Other less abundant epifauna included representatives of gastropod, decapod, polychaete, and pycnogonid taxa. Examination of patterns of epifaunal demonstrated that faunal-habitat associations differed significantly among macrophytes with divergent architectures. Moreover, these differences varied among epifaunal species with a disproportionately high abundance of epifauna on drift algae as compared to the seagrasses. While our study supported previous findings of a significantly higher abundance of crustaceans on macroalgae versus seagrasses, when examined at the species level, some pericarids displayed trends contrary to those observed in other studies. These overall findings reiterate that patterns of faunal abundance are probably maintained by a host of complex factors that extend beyond morphological features of the fauna and architecture of the macrophytes.

abstract

Figure 8.4. Main entry for the paper version of *Oceanic Abstracts*, April 1999, p. 50.

Copyright © 1999 *Cambridge Scientific Abstracts.* Reprinted with permission.

and it does not contain the complete content of the article, so it is always best to obtain the actual article if you plan to refer to it in your research.

Abstracting services frequently prepare separate, supplemental indexes in addition to subject headings and authors. *Oceanic Abstracts* includes a geographic index and a taxonomic index.

In figure 8.5, you can see that we could have located the sea grass article through the scientific (taxonomic) name of the organism discussed in this article *(Ruppia maritima)*. Many other abstracts provide specialized indexes that offer additional points of access to the literature in that field. You may find a chemical name index or an index that is arranged according to patent numbers. These specialized indexes provide us with more ways to find useful information.

Electronic Indexes and Abstracts

Electronic versions of indexing and abstracting services usually offer greater searching flexibility than printed sources. You may combine any keywords that you desire and you are not limited to predetermined subject headings and subdivisions. This can significantly improve your search output and relevancy, but can also lead to some frustration if proper search techniques are not employed.

When using electronic resources, you don't have to shuffle back and forth between multiple volumes. Summaries (or abstracts) are included in the full record along with the bibliographic citation, and you can often search multiple years simultaneously.

Every publisher likes a unique design for its product. This means that although many of the basic principles of electronic database search techniques are the same, you may encounter search screens that look very different as you move from one resource to another. Many databases offer a choice of basic or advanced search screens that appeal to a wide range of research needs and skills. The search screens will look similar to the OPAC screens, which are illustrated in chapter 7.

The "search terms" box allows you to input appropriate terms or phrases that represent your topic. The "search fields" box allows you to select the type of search that you would like to perform. Refer to the

scientific name

Partial decontamination of rotifers with ultraviolet radiation: the effect
  of changes in the bacterial load and flora of rotifers on mortalities in
  start-feeding larval turbot 4335
Ruppia,
  Succession of the bivalve *Abra ovata* community in Sulaksky Bay,
  Caspian Sea 3720
Ruppia maritima,
  The influence of habitat structure in faunal-habitat associations in a
  Tampa Bay seagrass system, Florida 3672

S   *accession number*

Saccorhiza polyschides,
  Modern cool-water carbonates on a coastal platform of northern Brit-
    tany, France: Carbonate production in macrophytic systems and
    sedimentary dynamics of bioclastic facies 3577
Saccostrea commercialis,
  Third generation evaluation of Sydney rock oyster *Saccostrea com-
    mercialis* (Iredale and Roughley) breeding lines 4333
Saccostrea echinata,
  Hatchery rearing of the tropical blacklip oyster *Saccostrea echinata*
    (Quoy and Gaimard) 4322
Sacculina carcini,
  Increased susceptibility of recently moulted *Carcinus maenas* (L.) to
    attack by the parasitic barnacle *Sacculina carcini* Thompson 1836 3631
Sacculina polygenea,
  Asexual reproduction as part of the life cycle in *Sacculina polygenea*
    (Cirripedia: Rhizocephala: Sacculinidae) 3533
Sagitta gazellae,
  Trophic importance of the chaetognaths *Eukrohnia hamata* and
    Sagitta gazellae in the pelagic system of the Prince Edward Islands
    (Southern Ocean) 3564
Salaria fluviatilis,
  Comparison of salinity tolerance and osmoregulation in two closely
    related species of blennies from different habitats 3479
Salaria pavo,
  Comparison of salinity tolerance and osmoregulation in two closely
    related species of blennies from different habitats 3479
Salmo salar,
  A survey of product defects in Tasmanian Atlantic salmon (*Salmo
    salar*) 4325
  Allozyme heterozygosity and development in Atlantic salmon, *Salmo
    salar* 3478

**Figure 8.5. Excerpt from the taxonomic index
of *Oceanic Abstracts*, April 1999, p. 275.**
Copyright © 1999 *Cambridge Scientific Abstracts.*
Reprinted with permission.

information in chapters 5 and 6 for suggestions on how to select appropriate keywords and fields necessary for performing an effective search.

An advanced search screen may provide additional boxes for the entry of multiple search terms and field searching options. In many databases, the field searching options available will be listed in a "pull-down" menu. Click on the arrow associated with the box to pull up the menu options, and then click on your choice of fields. Advanced search screens may also offer options to limit by date of publication or language.

In most databases, Boolean commands, proximity commands, and truncation may also be entered within the search term boxes to allow for quite sophisticated and efficient searches. The important thing to remember is that, although the design or layout of the search screen may vary greatly, the search principles usually do not.

Finally, regardless of whether you are using a printed or an electronic index or abstracting service, once you find a citation that looks interesting, it is *strongly recommended* that you copy down the *complete* citation exactly as it appears in the source. This information will be needed to locate the complete article on the periodical shelves or to request a photocopy of the article on interlibrary loan. Copying the citations may seem like a lot of work now, but it will save you headaches in the future!

Citation Indexes

The Institute for Scientific Information (ISI) publishes several specialized indexes called *citation indexes*. Citation indexes are unique among guides to literature. Like traditional indexes, they organize citations to periodicals, books, and reports and may be searched by keyword, author, or institution. As with any periodical index, the citation indexes include the bibliographic information you will need to find the full-text of the materials. These indexes depart from the traditional by also including a list of "cited references," or materials used by the author in the preparation of his or her book or article. These cited references can also be searched. The number of times a work has been cited by other authors may be indicative of its value to that area of study. In chapter 11, we will discuss how citation indexes may be used to help evaluate the impact of an article.

If you are using the printed version, the ISI Citation Indexes are issued in several parts: *Science Citation Index, Social Sciences Citation Index*, and *Arts & Humanities Citation Index*. The electronic version of the citation indexes is called *Web of Science*. In some cases, your library may subscribe to an electronic version that is issued from another *vendor*. In this case, the vendor receives the information from ISI and distributes its own *search engine*.

The initial screens for *Web of Science* offer you two different ways to search the database: a general search or a *cited reference search*. The general search allows you to input an author's name or keywords. You may also search the database by journal title or the author's institutional affiliation.

The searching method for *Web of Science* is quite specific. First names are not used in the author search. Each author entry includes only the last name of the author and, if known, the first and middle initials. If the middle initial is not given, you may replace it with an asterisk, which will act as a wildcard. If you want to search by the name of the publication where the article appeared, you must enter the publication name in a particular manner. A list of authorized entries for the periodical name is available via a *hot link* on the search screen. In chapter 11, we will illustrate a *Web of Science* search as we evaluate a specific article that appears in the journal *Marine Chemistry*.

You're Almost Home

The next step is to find the actual articles. First, determine whether your library subscribes to the periodical titles that you need. Go to your library catalog and perform a title search on the periodical title—not the title of the article. If the periodical is available locally, check the holdings to find out if the particular issue you need is available. The holdings section of the on-line catalog tells you the issues your library actually owns, if they don't have a complete run of the title. If the article is not available in your library, you might be able to obtain the material you need through an interlibrary loan or document delivery service.

Don't assume that if the title is not on the shelf, the library doesn't have it. In order to conserve space, some libraries retain older issues of periodicals on microform. Two common types of *microform* are *microfilm* (reels) and *microfiche* (individual sheets approximately four inches by six inches). In other cases, the title may be available electronically.

Again, the library catalog and/or the reference librarian should be able to confirm the holdings of a periodical title.

Locating Information from a Conference or Symposium

Many conferences and symposia publish abstracts or proceedings of the meeting. Information presented at scientific conferences and symposia may be written up as journal articles eventually. Much of it, however, never does, and it is often useful to get access to the content of the conference papers long before they go through the lengthy journal publication process. Engineers, in particular, often publish their results in *conference proceedings*, and the results never appear in a journal.

For recent meetings, you may be able to locate the proceedings on the *World Wide Web*. If not, you can look for the proceedings in a number of places. Many scientific indexes and abstracting services include conference proceedings. Some examples are *AGRICOLA, Aquatic Sciences and Fisheries Abstracts (ASFA)*, and *Engineering Index* (or its electronic version, *Compendex)*. You may also have access to specialized databases, such as *PapersFirst* or *ProceedingsFirst* or *Conference Papers Index,* which are specifically designed to index the content of conference proceedings. The actual process of searching an index for articles or abstracts in conference proceedings is exactly like that described for finding articles in journals. Many electronic periodical indexes also allow you to limit your search just to conference proceedings or just to journal articles.

Figure 8.6 shows the full record for a paper that was presented at the *First International Symposium on Flatfish Ecology*. The index record taken from *Aquatic Sciences and Fisheries Abstracts* provides information on the availability of the symposium proceedings. By examining the source (SO) field, we can see that, in this case, the proceedings were published as a special volume of the *Netherlands Journal of Sea Research*. Libraries owning the proceedings for the conference may decide to catalog the special issue of the journal like a book and put it into the circulating collection. Other libraries may shelve the proceedings in the periodicals section along with the other issues of the *Netherlands Journal of Sea Research*.

TI: Title
      Life history cycles in flatfish from the northwestern Pacific, with particular reference to their early
      life histories.
AU: Author
      Minami, T; Tanaka, M
AF: Author Affiliation
      Hokkaido Natl. Fish. Res. Inst., Katsurakoi 116, Kushiro, Hokkaido 085, Japan
CF: Conference
      1. Int. Symp. on Flatfish Ecology, Texel (Netherlands), 19-23 Nov 1990
SO: Source
      PROCEEDINGS OF THE FIRST INTERNATIONAL SYMPOSIUM ON FLATFISH
      ECOLOGY. PART 2., 1992, pp. 35-48, NETH. J. SEA RES., vol. 29, no. 1-3

AB: Abstract
      Nearly 300 species of Pleuronectiformes are found in the western Pacific, of which 120 species
      are recorded around Japan. In this area the geographical distribution of flatfish is generally related
      to family. Thus two major groupings are apparent; those from cold water, principally the major
      members of the Pleuronectidae, and those from warm-water habitats including members of the
      Citharidae, Paralichthyidae, Bothidae, Soleidae and Cynoglossidae. Notably, the family
      Scophthalmidae is not found in the northwestern Pacific. Japan's temperate location between
      subarctic and subtropical zones results in an overlap of the two groups. Most species inhabit the
      continental shelf, but some occupy waters deeper than 200 m and a few are found below 1000 m.
      Spawning occurs throughout the year at low latitudes, but tends to concentrate in summer at high
      latitudes. Flatfish tend to move to shallow water to spawn. Nursery grounds for juveniles are
      shallower than those of the spawning grounds.
LA: Language
      English
SL: Summary Language
      English
PY: Publication Year
      1992
PT: Publication Type
      Book Monograph; Conference
DE: Descriptors
      life history; INW, Japan; Pleuronectiformes; ecological distribution; spawning seasons; spawning
      grounds; vertical distribution; diets; Pacific Ocean, Northwest
ER: Environmental Regime
      Marine
CL: Classification
      Q1 01344 Reproduction and development; D 04668 Fish
SF: Subfile
      ASFA 1: Biological Sciences & Living Resources; Ecology Abstracts
AN: Accession Number
      2798885

Figure 8.6. Main entry for a symposium from the electronic version of
Aquatic Sciences and Fisheries Abstracts.

Once you locate a citation to a paper from a conference, you will need to determine if your library owns a copy of the proceedings. Conference proceedings are sometimes tricky to locate in your library catalog. Try combining several of the words from the conference name into a keyword search. In many cases, you can perform an author search on the name of the conference. Leave out the number or year, the frequency (e.g., annual), and the word "proceedings" before checking the catalog. For instance, the *Proceedings of the Twelfth International Seaweed Symposium* can be located by performing an author search on "international seaweed symposium." If you still aren't sure whether your library owns a copy of the proceedings, ask a reference librarian to help you search the catalog.

There are many, many conferences, and it is difficult for libraries to keep track of and acquire proceedings from them all. If your library does not own the proceedings you need, you may be able to obtain it on interlibrary loan. Libraries often treat conference proceedings like journals in that they won't allow them to circulate out of the library, but if you can provide your interlibrary loan librarian with a citation for the *specific* conference paper that you need, they can usually obtain a photocopy of that paper for you from another library or from a document supplier.

Getting the Latest News

Newspaper articles convey all kinds of information. There probably isn't one subject that hasn't been covered by newspapers in the history of news reporting, including science and medical news; and you can't beat the newspaper when it comes to timeliness. News articles often provide a firsthand account of breaking news, as well as recall past events, indicate trends, and prompt speculation, even before the weekly magazines hit the stand.

As great as these resources may be, we must offer a few words of caution. Newspaper articles may reflect the biases of the publisher, editorial staff, or the journalists. As with any resource, evaluating the information gleaned from a newspaper is very important. This is especially true of news articles that are reporting medical or scientific breakthroughs.

Another drawback is that news literature may not be indexed well, if at all. In the past, only a few of the more famous newspapers, such as the *New York Times*, provided a print index to articles appearing in their pages. Organized under broad subject areas, the citation entries are listed *chronologically* within each subject, emphasizing the importance of timeliness to these publications. In most cases, the author and headline are *not* included in the citation. A brief summary of the article is sometimes included followed by an indication of the length of the article: S (short), M (medium), or L (long). The month, day, newspaper section, page, and column are also identified, in that order. When using a print newspaper index, it's important to note the year(s) covered by the index.

To locate information in newspapers that were not indexed, you either had to know a specific date or look forward to *long* hours of browsing paper or microform copies. By the late 1980s, however, many newspapers were automated. These electronic databases not only indexed the newspaper, but made the full-text content of the articles available electronically as well.

Today, general and subject-specific print and electronic indexes may include coverage of news literature as well as journals or magazines. Libraries may subscribe to individual electronic newspaper titles or to an *aggregator* service such as *Newsbank* or *Lexis-Nexis*. Aggregator services provide access to full-text articles from a large number of newspapers and journals, and the information is searched as one large database. One drawback of print newspaper indexes is the time needed to accumulate a month or so of newspapers and print and mail the indexes, partially negating the timeliness of newspaper information. Many of the electronic newspapers and aggregators are available within twenty-four hours of the publication of the newspaper. Because of the vast size of the databases, these full-text resources require advanced searching techniques as mentioned in chapter 6. Using Boolean commands and field searching, and limiting your search by date, can really help to focus a search. The downside to the electronic versions of newspapers is that many do not yet contain graphics. If a photo, a chart, or a map is needed, you still may have to obtain a print or microform copy of the newspaper.

Summary

Periodical articles may be desirable for their timeliness and conciseness. Whether the periodical index or abstracting service you are using is in the printed or electronic form, there are several things you will want to keep in mind:

- Select an appropriate index for your subject area. If you are unfamiliar with the index, take a few minutes to familiarize yourself with it by reading the preface or by scanning through the help screens.
- Think about what keywords or subject headings will be most useful for your search, given the index you are using.
- When you locate article citations that appear useful for your research, be sure to copy down the complete citation as it is listed in the index.
- Determine if your library owns the journals that you need or if you will be able to request the materials through your interlibrary loan department or a document delivery service.

Ulrich's International Periodical Directory is a useful tool for gaining information about journals and magazines. It is *not* an index to periodical articles.

Citations to conference proceedings may be included in periodical indexes or abstracting services. To determine if your library owns a particular proceedings title, try a keyword or author search on the name of the conference.

Newspaper articles can provide timely, firsthand observations on a particular topic. A few things to remember about printed newspaper indexes include:

- Printed newspaper indexes are arranged chronologically within each subject area.
- Be sure to record important citation information such as year, date, section, page, and column in order to locate the newspaper article.
- Print indexes are produced only for major newspapers, but electronic aggregators are now available that provide timely access to the full-text of many more newspapers.

Chapter 9

Locating Quality Information on the World Wide Web

In the not too distant past, discussions of library research centered on methods that a student could use to find adequate amounts of information to write a term paper. The Internet has changed all that by allowing the research community the ability to disseminate an amazing amount of information quickly and easily. Now the emphasis in library instruction tends to be on how to cull through the massive amounts of information that are normally retrieved in a simple search.

The Internet and the World Wide Web

People have a tendency to use the terms Internet and World Wide Web interchangeably; however, they are not exactly the same thing. The Internet is a communication network that allows computers around the world to talk to one another using a common language or *protocol*. There are a number of services that utilize this communication network to disseminate information. Among other things, the Internet may be used for electronic mail, electronic discussion lists, file transfers (FTP), or for displaying information that is retrieved from another computer. The World Wide Web is a method of information display that uses the communication abilities of the Internet along with a software program called a *browser* to access graphical and multimedia images stored on one computer *server* and then display them onto another computer. *Netscape Navigator*, *Netscape Communicator*, and *Internet Explorer* are examples of Web browsers that allow your computer to interpret the messages that it receives over the Internet into a format that you can read on the screen.

There are entire books that have written about the Internet and the World Wide Web that explain these concepts in detail. It is beyond the scope of this book to provide that kind of depth. Our goal is to make the reader familiar with some of the terminology used in Web searching, and to provide tips on how to locate quality Web sites and improve your searching techniques.

Free Web Sites versus Proprietary Databases

A great deal of this book is centered on searching for information through the use of electronic databases. What many people may not realize is that the commercial indexes and many of the full-text databases that we have discussed in other chapters utilize the Internet to transfer data. A library purchases a subscription that allows its users access to a publisher's database.

When the contract is negotiated for a particular resource, it normally includes language that identifies authorized users who may have access to the Web site containing the data. Institutions have assigned *IP addresses* that are used to identify machines within an institution. Many publishers use these IP addresses to control access to their Web sites. If the computer that you are using has an IP address that is authorized to access the publisher's site, you may perform your searches. If the IP address falls outside the IP address range specified in the contract, you will get a message telling you that you may not access the site. This is why it can be difficult to obtain access to a library database from your home computer. If you use a commercial *Internet Service Provider (ISP)* to access the Internet, the publisher's Web site cannot immediately recognize you (through your computer) as a person who is authorized to access their site since your computer is not registered with the publisher.

Some institutions have set up a *proxy server* system to allow their users remote access to the databases that the institution has purchased. With a proxy server, you configure your Web browser so that it asks you for identification. Most institutions ask for a faculty or student identification number as proof that you are currently affiliated with that institution. The identification number travels over the Internet to a proxy server located at your institution. The proxy server compares the number to the list of authorized users in its database and, if the number is present, allows you to access the publisher's Web site.

The rest of the discussion in this chapter will focus on locating Web sites that are not part of a commercial subscription—sites that can be accessed by anyone for free without having to pass through an authorization process. We will refer to this as searching the "open" Internet. Purchased Web sites have already passed through a certain level of selection. The publishers have reviewed the information prior to adding it to their sites. In addition, before a library decides to purchase a subscription, an electronic database normally passes through a fairly extensive review by a librarian or team of librarians in charge of deciding for what materials they will pay to make them available to their users.

On the other hand, the *open Internet* contains information that is not only available to anyone; anyone may publish or make their work available on the Internet. The author of the Web site you are accessing may be a renowned scholar, or it may be a teenager who is looking for ways to fill his time. It isn't always easy to determine who has authored a Web site, so you need to take great care to ensure that the site is reputable. In chapter 11, we will provide a number of tips on how to evaluate a Web site, but first we need to locate the sites.

Search Engines and Directories

As we have mentioned, there is an overwhelming amount of information available through the World Wide Web. One method that has been used to help organize the open Internet is the creation of directories, such as *Yahoo*. These provide a hierarchical method for locating Web sites. You begin with very general topics such as "Education," "Health," or "Science." If you click on the Science heading, you will be offered several choices that are somewhat more specific than the broad topic of Science. At this stage, your options might be, "Biology," "Chemistry," "Engineering," and "Geology." Clicking on any of these topics will give you a list of narrower subjects that relate to the main heading. As you work through the lists, you will eventually reach a list of Web sites on the most specific topic. Directories can be useful for beginning Web searchers since they lead the researcher through the process; however, experienced searchers may find them somewhat restrictive and time-consuming. Directories do have the advantage in that human beings compile them, and the sites listed have undergone at least some level of evaluation before being listed.

Search engines may also be used to look for information on the open Internet. A search engine is a software program that has the ability to scan the World Wide Web for sites. It then allows you to enter key terms and the search engine will look for sites within its database that contain the search terms you entered into the program. You may be familiar with various search engines such as *AltaVista*, *HotBot*, or *Excite*. There are many more released every day, but these are a few of the more familiar search engines.

When you perform a search on the Internet using a search engine, your results will normally vary according to the search engine that you use. There are several reasons for this, but the primary reason is that each search engine is scanning a different database of Web sites. Search engines employ *crawlers* or *spiders* to locate new and updated Web sites for their databases. A spider is computer software that is programmed to scan the Web on a regular basis. As new Web sites are encountered, the site is added to the search engine's database and the Web site becomes available for retrieval from that search engine. Since the spiders scan the Web at different times and at different intervals, search engines vary in their size and completeness.

Another way that your results may vary is in the search method that is employed by the search engine. Some search engines index the title of the home page and the words that appear on the home page. Others may look at every word that is available on the Web site. In some cases, the search engine will return one result for each Web page in a *domain*. Other search engines may provide multiple results for a single Web page by giving a result for every level of a Web page that appears relevant.

Additionally, the sites that you retrieve from a search engine will vary because of the method of relevancy ranking that is used. Search engines normally list the results of a search in relevant order—that is, the Web sites that contain all the search words, or contain them most frequently, or contain them in the most prominent places are listed first in the results list. Different search engines may employ slightly different methods for ranking results, which can also lead to inconsistencies in retrieval.

Tips for Better Internet Searching

When you are ready to perform a search of the open Internet, you should begin by selecting a search engine such as *HotBot, Google, Lycos,* or *AltaVista.* Try out different search engines to see which ones contain features that you feel are most useful. Compare the results of the same search using a couple of different search engines to see which ones retrieve sites that are most helpful to your own research. Some search engines, such as *Dogpile,* now perform the same search in multiple search engines, allowing you to compare the results directly. Ask your friends and colleagues which search engines they prefer. With experience, you will probably find yourself tending towards one or two favorite search engines for your research. This is perfectly fine, since it gives you the opportunity to learn the advanced searching methods that help to refine your search. If you use one of your favorite search engines and don't find anything worthwhile for a particular topic, you can always try other search engines to see what you might find.

To make the searching system appear neat and easy, most search engines default to a very basic search screen that may consist of a single search box. Although you may have to look carefully to find the link, nearly all of the well-known search engines provide advanced searching techniques such as Boolean and proximity searching that will allow you to perform more focused searches. Although the techniques remain the same, the commands that are used differ from one search engine to another and some are more user-friendly than others. Some search engines will actually walk the user through a search. They may provide a single entry box for search terms, but then give you the option to search for all the words, any of the words, or for the exact phrase that you have entered. In this case, you don't have to memorize any Boolean or proximity commands. With some search engines, you can perform very flexible and complicated advanced searches, but you have to know the exact format for the Boolean, truncation, and proximity commands. As with any database, you can locate tips on how to search a particular site through its help options. Since a typical open Internet search retrieves tens of thousands of hits, this is an area that cries out for the use of advanced searching techniques.

Selecting Keywords

As we have mentioned numerous times throughout this book, selecting useful search terms is one of the most important steps in electronic searching. This is particularly the case when it comes to open Internet searches because the size of the database is so large. Always be as specific as you possibly can when you choose the keywords you want to enter into an Internet search. If you want to learn about algae in Monterey Bay, use "algae" and "Monterey Bay" as your keywords; don't search for just "algae," or even for "algae" and "California." If you are interested in a particular species of algae in Monterey Bay, use the species name. This can serve two purposes: to focus your search even more and to help eliminate the less scientific Web pages.

Case Sensitivity

Case sensitivity refers to the ability of the search software to differentiate between capital *(upper case)* and lower case letters. In the early days of Web searching, it was very important to be case specific when performing a search. Many of the search engines are now a bit more forgiving of case-sensitive searches—a search for a word typed in *lower case* letters may retrieve any form of the word whether it is capitalized or not. It pays to check out the treatment of case sensitivity, though, to be sure that you know how the system is interpreting your search statement. It is important to know if you must capitalize proper nouns in order to pull up the results that you are seeking.

Boolean Searching

Boolean searching should be available in any of the popular Web search engines; however, the format can vary even more than it does in purchased or proprietary databases. As you may recall from chapter 5, most databases use the AND, OR, and NOT formats when combining search terms. In some cases, the Internet search engines use this same format. It is also not uncommon for them to use the *unary system* where a plus sign is similar to the Boolean AND and the minus sign is similar to the Boolean NOT command. Preceding a search term with a plus sign means that a search term must be present in the final results. Preceding a keyword with a minus sign excludes that particular term from the results.

A search for: "+algae +Monterey" tells the search software that both terms, algae and Monterey, must be present in the Web pages that are retrieved. A search for "+algae +Monterey -Aquarium" will instruct the search engine to look for Web pages that include the words algae and Monterey but exclude those that contain the term Aquarium. Web pages related to the Monterey Aquarium would not be included in the results.

As with all the specialized searching techniques, it may be necessary to check the help screens to find out how Boolean searching is conducted. While some search engines use regular Boolean commands (AND, OR, NOT), and others use the plus and minus system, it is also common for an Internet search engine to employ a pull-down menu that emulates Boolean combinations. In this format, a search for "all the words" would be similar to a Boolean AND search. A search for "any of the words" corresponds to a Boolean OR search. Finally, it may be possible that different searching techniques may be available within the same search engine. The basic search screen may employ the pull-down menus while the advanced screen may allow the entry of AND, OR, and NOT Boolean commands.

A final word of caution: always check the help screens to determine the default search mode. Many people do not include Boolean commands when they enter keywords into a basic Internet search screen. That is fine, as long as you know how the search will be performed. Will the search engine combine those terms using a Boolean AND search so that all terms must be present in the final result? Or, will the software treat the search as a Boolean OR search so that any, but not necessarily all, of the keywords will be present in the results? The final search results can vary greatly, so it pays to know how your search results were constructed.

Proximity and Phrase Searching

Another useful way to focus your search is through the use of proximity and/or phrase search commands. As you recall from chapter 6, proximity searching is employed when the computer software is instructed to search for one word within a specified distance of another word. Some Internet search engines allow proximity searching, especially in their advanced search forms. Check the help screens to find out what proximity commands are allowed and how they are used.

Phrase searching, a very specific type of proximity searching, is another useful way to reduce your search results to more relevant Web pages. In a phrase search, the computer software is instructed to look for two particular words right next to one another and in the order in which they are typed. A phrase search of "Shell Island" should be much more relevant than a Boolean search for "shell" and "island." In the phrase search, you retrieve Web pages that discuss a specific geographic location called Shell Island. The Boolean search will provide you with those Web sites but will also retrieve many, many Web pages that discuss shells on islands—any shells on any island, not just Shell Island.

Limiting

Another advanced searching technique that was introduced in chapter 6 is the concept of limiting your results to a particular field or time period. Internet search engines often include a variety of limits that can be used in the advanced search mode to focus your search.

A date limit is often available that allows you to focus your search on Web pages that were published within a certain time frame. In many cases you can select between predetermined time frames (previous week, previous month, previous year, etc.) or you can customize your own date range. This can help to limit access to older Web sites that may not be as relevant to your research. Or, conversely, it can be used to limit your search to Web sites that have been around for a while and, therefore, may be more stable sites.

Another popular limit is the ability to restrict the search to the title of the Web page. In this case, the link to the Web page will only be retrieved if the keyword is included in the title of the page. This can significantly reduce your retrieval and pull up only those pages where the keyword was considered important enough to be included in the title.

Most search engines also allow you to select a language limit. Since the Internet is a global network, it is possible that you will retrieve Web pages that include a significant amount of non-English text. If you are not interested in examining Web sites that are printed in a foreign language, you can use the language option to limit your search to English only text. This feature can be even more useful if you want to search for a foreign language site.

Perhaps you need to locate a moving image of a tornado or an audio file that will allow you to listen to the call of the Great Horned Owl. This kind of information is quite accessible if you add a limit for media to your regular keyword search. Keep in mind that you may need to add additional software to your computer if you want to see or hear a multimedia file.

Restricting searches to certain types of Internet domains may be valuable. The domain is a part of the Internet address that is assigned to a Web page based on the type of institution or author that has produced the site. Academic institutions in the United States have an ".edu" domain attached to their Web address, United States government sites have a ".gov" domain, commercial sites a ".com" domain, and non-profit organizations have a ".org" domain. These are a few of the more common domains that you will see as you search for sites that are authored in the United States; however, there are many additional domain codes that are available. If you are interested in finding information on research that has been done on a particular chemical compound, you might want to limit your search to the .edu or .gov domains. This will focus your search on materials that have been produced at colleges and universities or at government-sponsored sites. It will also remove any sites that are discussing the chemical in a commercial sense.

Check the advanced search help screens to determine other ways that you can use limits to focus your search.

Wildcards and Truncation

Some of the Internet search engines do support wildcard searching where a symbol is used to replace one or more letters. This should be used with great care when searching the Internet since it tends to greatly expand your search results and, in most cases, you will want to narrow rather than expand your search. The help screens will tell you what symbol is used for truncation if it is available in the search engine that you are using.

Beyond Search Engines

This chapter has concentrated on techniques to improve open searching of the Internet using a Web search engine. There are additional ways that you can locate relevant Web sites for your research.

Several commercial vendors are now scanning Web sites and adding links to them from their databases. The *FirstSearch* service has a separate database called *NetFirst* that is devoted entirely to Internet resources. The advantage to using a database like *NetFirst* is that it contains Internet sites that have been reviewed and organized for easier research. Catalogers provide abstracts and subject headings to Internet sources that they feel will be useful to researchers. In addition, the sites listed in *NetFirst* are regularly checked to be sure that they are still active. *NetFirst* can be a good place to start when you are hoping to find stable, useful Web sites.

Another vendor, *Cambridge Scientific Abstracts*, is now also providing links to Web sites. If you perform a search in any *Cambridge Scientific* database, your results may include a link to Web sites that appear to be related to your search. A *Cambridge Scientific* editor has entered every site listed in the Web Resources Database, and the choices must meet specified criteria before they are considered for inclusion. *Cambridge Scientific* also checks its Web sites regularly to ensure that the sites are active.

It is also becoming much more common for bibliographies to include Web sites. Check the list of references in a recent journal article and, chances are, you will see some Internet sources included in the list. Some books and journals are also including *"webliographies"* or bibliographies that concentrate on Web sources. These resources can be particularly helpful since they have been evaluated by the author and are considered useful.

Another excellent resource for "free" Web sites may be as close as your own library's Web site. Many academic and public libraries are creating their own virtual libraries that contain carefully selected Web sites. Subject specialist librarians scan the literature and the Web to locate sites that they, or other experts, feel are valuable to research.

Evaluating Your Search Results

Printed books and journals go through a preliminary evaluation process when editors and peer reviewers accept them for publication. Unless a book is published by a *vanity press*, you can assume that it has at least a minimal level of credibility. The open Internet is another story entirely. Anyone with access to a computer and an Internet provider has the capability of "publishing" on the Web.

Critical evaluation of the material that you retrieve over the open Internet is absolutely essential if you are planning to use the data in your research. Chapter 11 provides numerous tips for analyzing the credibility of a Web site.

The Future of the Web

It is difficult to even begin to predict what the World Wide Web will be like next year, much less several years from now. Web search engines have improved drastically during the last few years and, hopefully, as more and more people sign on, it should become even more organized.

Summary

The Internet contains a tremendous amount of useful information and a great deal of useless information. Careful searching can help to focus your Web search to more useful sites.

One of the most important steps in a good search of the Internet is the selection of appropriate keywords. Because the amount of information is so immense, it is important to be as specific as you possibly can in the selection of keywords.

Advanced search techniques are available in most Web search engines. Check the help pages to determine the availability and format for:

- Boolean searches
- Proximity and phrase searches
- Limits (date, title page, language, media, domain, etc.)
- Wildcards and truncation

Web directories such as *Yahoo* and search engines such as *Google* and *HotBot* do not provide the only method for locating useful Web sites. Webliographies and your campus library's Web site may also have links to pertinent sites. Purchased databases such as *NetFirst* and *Cambridge Scientific Abstracts* also include focused ways to access current, relevant Web sites.

Chapter 10

Government Resources

A great deal of scientific research is funded by, or in some way affiliated with, an international, federal, state or regional government agency. Government agencies publish all kinds of information: technical reports, statistical information, maps, government hearings and regulations, and much, much more. In this chapter, we concentrate mostly on documents published by the United States government. These documents are widely cited in scientific research and, in most cases, relatively easy to locate. Tips for locating other types of government documents may be obtained from your reference librarian.

The Government Printing Office (GPO)

The United States *Government Printing Office (GPO)* is one of the largest publishers in the world. The GPO was established in 1860 to handle the publishing needs of the United States Congress. Since then, the GPO has expanded tremendously, and it is now responsible for meeting the publishing needs of the majority of the United States government departments and agencies.

The Government Printing Office does much more than print documents. It is also responsible for cataloging and disseminating government information in printed and electronic formats. Government information is distributed in a variety of ways: it may be distributed to *depository libraries*, it may be disseminated over the Internet, or it may be sold to the consumer.

Depository Libraries

The Federal Depository Library Program began in 1813 as a method for providing free access to government information. Federal depository libraries apply for depository status and are designated as such by members of the United States Congress. Depository libraries receive many government documents free of charge and are required to follow strict guidelines in order to retain their depository status. Foremost among these requirements is that they must provide free, open access to the public and have knowledgeable staff available to assist the users.

There are more than 1300 depository libraries in the United States, and many of them are located in college and university libraries. Most depository libraries are categorized as *selective depositories*, meaning that they receive only a portion of the depository items that are available to them. Each state also has, or has access to, a regional depository library. *Regional depositories* are responsible for collecting all the depository items that are available from the government. Libraries that have not been selected as depository libraries may still add government documents to their collections; however, they normally have to purchase the documents from the GPO. To locate the depository library closest to you, ask your reference librarian or access the GPO Web site on the Internet.

Electronic Access to Government Information

In 1993, the Government Printing Office Electronic Information Enhancement Act was passed to provide for the free, electronic dissemination of government information. This law led to the formation of *GPO Access*, which was introduced in 1994. *GPO Access* is available on the Web at http://www.access.gpo.gov. *GPO Access* includes links to more than 1500 databases including, the *Code of Federal Regulations*, the *Congressional Record*, *Economic Indicators*, the *Federal Register* and the *United States Code*. The GPO is responsible for making sure that the information contained in *GPO Access* is secure. The text, graphics, and other content may not be changed except for official changes submitted by the GPO.

Purchasing Government Documents

It is also possible to order documents directly from the Government Printing Office. The *GPO Access* Web site has a link to an on-line bookstore, which allows private consumers to purchase documents through a variety of methods: over the Internet, by telephone, by fax, through the mail, or by teletype. Specific instructions on how to order materials are listed on the United States Government on-line bookstore Web site at http://bookstore.gpo.gov/index.html.

In addition, there are a number of GPO bookstores located throughout the United States that have documents in stock or can place orders for GPO materials. A listing of the locations of these bookstores can be found in the *United States Government Manual* or at http://bookstore.gpo.gov/locations/index.html.

The Superintendent of Documents (SuDocs) Classification System

United States government documents are organized in depository libraries according to the Superintendent of Documents (SuDocs) Classification System. A typical *SuDocs number* would look like this: C55.2:H 94/8. Similar to the Library of Congress Classification system, which was discussed in chapter 7, the SuDocs Classification System at first glance may look like an incomprehensible string of letters and numbers. The SuDocs system is actually quite easy to understand once you know how it is formulated.

Unlike many library classification systems, which arrange items by subject, the SuDocs system classifies documents by issuing agency. This means that if you are looking for documents on a particular subject you may find them in a number of different places in the stacks. Suppose that you want to find government information on hurricanes. You might locate documents in the National Oceanic and Atmospheric Administration (NOAA) section that discusses how to track hurricanes, C 55.122:H 94/SMALL. Another document in the Department of Housing and Urban Development section may tell about hurricane damage to people's homes, HH 1.2:B 43/2. Or a U.S. Geological Survey report may provide information on the geological and hydrological effects of hurricanes, I 19.42/4:00-4093. Yet another area may contain

Congressional hearings related to federal disaster relief, Y4. IN8/16:R 27/12. Each one of the documents will be shelved with other documents from the issuing agency so they may be located a great distance from one another in the stacks. This system is useful for finding information from a particular agency; however, it makes it difficult to browse the stacks for a specific topic.

As we mentioned earlier, the first consideration for any government document is its parent agency within the U.S. government. The first set of letters in a SuDocs number refers to the issuing agency. The parent agency or parent organization is always assigned the number one. Subordinate departments within the parent agency are assigned the same letter, with higher numbers. For example, the U.S. Department of the Interior has been assigned the letter "I." If the Department of the Interior produces a publication, it will be assigned the classification "I 1." Agencies within that organization would have the letter I with a specifically assigned number that is higher than the number one. It would look something like this:

> I 1   U.S. Department of the Interior
> I 19 Geological Survey (USGS)
> I 27 Bureau of Reclamation
> I 29 National Park Service
> I 49 Fish and Wildlife Service
> I 53 Bureau of Land Management

The USGS, the National Park Service, the Fish and Wildlife Service, the Bureau of Reclamation, and the Bureau of Land Management are all bureaus or agencies that answer to the U.S. Department of the Interior.

The agency designation is followed by a period and then a number that indicates the *type* of publication that is being classified. Again, these follow a logical order:

> 1   annual report
> 2   general publication
> 3   bulletin
> 4   circular
> 5   laws
> 6   regulations, rules and instructions
> 7   releases
> 8   handbooks, manuals, guides

9	bibliographies and lists of publications
10	directories
11	maps and charts
12	posters
13	forms
14	addresses, lectures, etc.

The numbers 1-14 have the same meaning for all agencies. Numbers above 14 are used for specific series that don't fit into the first 14 classes. This first part of the SuDocs classification number is referred to as the *class stem*. A colon completes the class stem. Following the colon is the book number, which further identifies the item by representing the title or author of the work. The book number may also include information relating to a report number, series number, volume, or date of publication.

Let's begin with a very basic example. The Bureau of Land Management published a map in 1980 that was titled *Wilderness Review, Michigan*. It was assigned the SuDocs number: I 53.11:M58. First we will examine the class stem: I 53.11. The first section, I 53, stands for the U.S. Bureau of Land Management, which is part of the Department of the Interior. The next section, .11, refers to the type of publication: maps and charts. Finally, the book number follows the class stem: M58. This section is based on the word "Michigan," which was assigned to organize the map alphabetically amongst the other maps published by the Bureau of Land Management.

Now let's examine an item that is part of an ongoing publication. The U.S. Geological Survey publishes a series called the *Geological Survey Bulletin*. Each title in this series has a number. In this case, the report titled *The Geologic Story of Isle Royale National Park* was issued bulletin number 1309 in the series and assigned the SuDocs number: I 19.3:1309.

Let's look at how this was classification number was created. The first part of the SuDocs number, I 19, refers to the issuing agency, in this case the U.S. Geological Survey, which is part of the Department of the Interior. The next part of the classification number tells us the type of publication. In this case, .3 is the category for bulletins. The last part of the SuDocs number following the colon indicates that this is bulletin number 1309 of the *Geological Survey Bulletin* series.

Now let's examine one more, relatively complex example. The U.S. Fish and Wildlife Service published a document titled *Devils Lake Wetland Management District, North Dakota*. This document was assigned a SuDocs number of: I 49.2:D49/2/996. Again, the first section, I 49, refers to the issuing agency, U.S. Fish and Wildlife Management, which is part of the Department of the Interior. The number following the period, 2, is assigned to general publications. After the colon is the book number, D49, which is based on the word "Devils." All this is pretty straightforward. Notice, however, that at the end of the SuDocs number we see a /2. If more than one item has already used the I 49.2:D49 designation, additional documents may be classified using the same number. They are numbered sequentially as they are published. The /2 here refers to the second publication that received this SuDocs base number. Finally, the /996 at the end of the classification number tells us that this document was published in 1996.

Similar to the Library of Congress classification system, documents using the SuDocs system are filed on the shelf alphanumerically. For example, U.S. government documents published by various departments and agencies would appear in the following order on the shelves:

> A 43.16/2:C76
> I 49.2:AB3
> LC 5.2:R13
> LC 29.9:H18
> LC 30.2:B47/2/999
> LC 30.2:C17
> SI 13.2:Sm 5
> T 28.2:C66/11

Locating Government Documents

How can you figure out what publications have been issued by the United States government? If your library is a depository library, the publications that they have selected for their government holdings may be listed in your on-line catalog. Since a large number of government-sponsored publications are not attributed to a single author, you may find a corporate author search on the government agency name to be a useful way of locating these items. Refer to chapter 7 for more information on how to conduct a corporate author search. If your library is not a depository library, or if you want to know what kind of publications exist that are not available at your library, you have several additional options.

The *Monthly Catalog of United States Government Publications* is considered the primary index to government documents. The printed edition of the *Monthly Catalog* began publication in 1895 and has seen numerous changes in the years since.

The print version of the *Monthly Catalog* is arranged much like a print abstracting service in that you normally begin your search by using one of several indexes. The indexes lead you to a citation number. The full record is listed numerically according to the citation number. Although the selection of indexes has varied over the years, most editions of the *Monthly Catalog* contain at least subject, author, and title indexes. The indexes may be separated or they may be interfiled alphabetically into one index.

The subject index is of particular importance since the full citations are listed according to the authoring agency. As mentioned earlier, documents on a similar topic may be published by many different government agencies so you will want to use the subject index to be sure that you have located all relevant materials.

Figure 10.1 illustrates a typical *Monthly Catalog* entry. We can see that the accession (citation number) is 87-18698. These numbers are arranged numerically throughout the volume, with the "87" designation simply meaning that the citation is located in the 1987 volume. The title of the document is *A revision of the neotropical aquatic beetle genera Disersus, Pseudodisersus, and Potamophilops (Coleptera: Elmidae)*. The document was written by Paul Spangler and Silvia Santiago and appeared as issue number 446 of the *Smithsonian Contributions to Zoology* series. To locate the document on the shelf, we would go to the SuDocs number, SI 1.27:446.

The middle section of the citation displayed in figure 10.1 provides additional information about the document. From this entry, we can see that the document is 40 pages long. It also tells us that it is illustrated and that it contains maps. It also contains a bibliography. Finally, the large black bullet ● in the entry indicates that this document was distributed for free to depository libraries, if they chose to select it for their collections.

The bottom part of the entry lists subject headings and additional materials that are used by the catalogers to describe this particular document. This section also provides both the Library of Congress and Dewey Decimal classification numbers for use by non-depository li-

citation number

SuDocs number

87-18598

SI 1.27:446

author ⟶ Spangler, Paul J.

A revision of the neotropical aquatic beetle genera Disersus, Pseudodisersus, and Potamophilops (Coleoptera: Elmidae) / Paul J. Spangler and Silvia Santiago. — Washington, D.C. : Smithsonian Institution Press, 1987.

document title

document is part of the Smithsonian Contributions to Zoology series

date of publication

publisher

document is 40 pages long, contains illustrations and maps ⟶ iii, 40 p. : ill., maps ; 28 cm. — (Smithsonian contributions to zoology ; no. 446) Shipping list no.: 87-481-P. Abstract in English and Spanish. Bibliography: p. 40. ●Item 910-D ⟵

indicates this item is available to depository libraries

subject headings ⟶ 1. Disersus — Latin America — Classification. 2. Pseudodisersus — Latin America — Classification. 3. Potamophilops — Latin America — Classification. 4. Insects, Aquatic — Latin America — Classification. 5. Insects — Latin America — Classification. I. Santiago, Silvia. II. Smithsonian Institution. Press. III. Title. IV. Series. QL1.S54 no. 446 86-600252 //r87 [QL596.E45] 591 s [595.76/45] /19 OCLC 14068944

Figure 10.1. *Monthly Catalog of United States Government Publications,* December 1987, p. 126.

braries that may have purchased the document for their regular collections.

The *Monthly Catalog* is also available in an electronic format. You may find it on CD-ROM or accessible through your library's Web page. Like many electronic indexes, however, the electronic version of the *Monthly Catalog* is usually much more limited in its coverage. Most electronic versions of *Monthly Catalog* index only documents that were published from 1976 onward. For more recent research, though, the electronic versions provide enhanced search features such as Boolean and proximity searching.

If, for some reason, your library does not have the *Monthly Catalog*, there are still ways to get access to government documents. Government documents are listed in the *WorldCat* database that is available through the *FirstSearch* service. This database is discussed in greater depth in chapter 7.

GPO Access, the Web site that was discussed earlier in this chapter, provides an equivalent edition of the *Monthly Catalog* called the *Catalog of United States Government Publications*. It contains indexing for documents published since 1994. The *Catalog* is updated daily and has a user-friendly search interface that permits Boolean and proximity searches. Researchers can search the database by keyword, title, or SuDocs class number. Best of all, the database is free and available to anyone who wants to use it.

National Technical Information Service (NTIS)

Not every publication that is issued by the federal government is available from the Government Printing Office. The National Technical Information Service, better known as NTIS, is an extremely important resource for scientists. This agency, which is a part of the U.S. Department of Commerce, acquires and sells information about unclassified, government-funded research projects in the areas of science, engineering, technology, and business. NTIS also collects information from a number of international sources. With more than two million publications on a wide variety of topics, NTIS is considered to be a primary source for government-related, scientific research documents.

There are several ways to locate information that is available from NTIS. As with many resources these days, your library may subscribe to a commercial version of the NTIS database. Access may be available either through the Web or on a CDROM. Dates of coverage for the electronic database vary depending on the subscription, but some vendors index NTIS materials as far back as 1964. You can also access a free version of the NTIS database at the NTIS Web site at http://www.ntis.gov. This resource covers reports that have been issued by NTIS since 1990. Your reference librarian can help you to determine the best way to access this information.

Remember that the NTIS index is not a full-text resource. A search in this resource will provide citations for reports that are available from NTIS. Once you have the citation, you can check your library's on-line catalog to determine if your library has purchased that particular report for its collection. Many of the NTIS reports are purchased on microfiche, so you may want to check with the reference librarian or a government documents specialist to see if the NTIS materials are kept in a separate area. If the report is not available at your local library you may attempt to obtain it through your interlibrary loan department. It is also possible to purchase many of the reports directly from NTIS. Information on how to check on product availability and the purchase price are included on the NTIS Web site. Keep in mind that the reports may be fairly costly so you will want to verify the price before you place your order.

Gray Literature

Gray literature refers to materials that have been published but have not been incorporated into traditional indexing services. Because of the lack of standardized indexing, these materials, which are often of great importance to research, may be extremely difficult to locate. If you find a citation for a published report that you have been unable to locate, ask the reference librarian or your interlibrary loan office for help. They often have some tricks of the trade that they can use to track down the report. The United States Department of Energy Office of Scientific and Technical Information (OSTI) recently started an endeavor aimed at locating specialized federal literature. This project, called *GrayLIT Network*, is available on the Web. It provides a method for cross searching several government Web sites. At the date of this publication, *GrayLIT Network* provided access to reports from the U.S. Department of Defense, the U.S. Department of Energy, the U.S. Environmental Protection Agency, and NASA. The *GrayLIT Network* is available at http://www.osti.gov/graylit/.

Statistical Information

Statistical facts can provide important support to a research theory. The United States government is also an excellent source for statistical information. Statistical reports are published by many government agencies and can be located using the *Monthly Catalog* or several other specialized resources, such as the *American Statistics Index*. As always, a reference librarian or a government documents specialist can help you to locate the indexing tools that will be most useful for your particular research topic.

If you are in need of basic statistical information, you may want to turn to the *Statistical Abstract of the United States*. This handy ready-reference volume, first published in 1878, contains an overwhelming amount of statistical information on all aspects of American life. Perhaps you want to locate the amount of electricity consumed in the residential sector of the United States. Or maybe it would be helpful to know the amount of research and development money that has been expended at colleges and universities on science and engineering projects. Ever wondered how many millions of dollars are spent on NASA space shuttle operations? All that and more are available in the *Statistical Abstract*. Statistical resources often contain an index to the tables. Be sure to look at the resource carefully. Sometimes the numbers listed in the index refer to table numbers rather than page numbers.

When using the *Statistical Abstract*, or resources like it, always be sure to check the explanatory material that is included with each table. This section will tell you what agency compiled the original data. It may be the U.S. Census Bureau, the U.S. National Science Foundation, NOAA, or any of a number of government agencies. It will also include any footnotes that help to clarify the statistics that are listed.

Many statistical tables conserve space by abbreviating the numbers in the table. The explanatory material will tell you if you should multiply the numbers listed by another number to come up with the actual statistic. Figure 10.2 illustrates a table from the *Statistical Abstract of*

No. 1156. Fishery Products—Domestic Catch, Imports, and Disposition: 1980 to 1998

[Live weight, in millions of pounds (11,357 represents 11,357,000,000). For data on commercial catch for selected countries, see Table 1381, Section 30, Comparative International Statistics]

Item	1980	1990	1992	1993	1994	1995	1996	1997	1998
Total	**11,357**	**16,349**	**16,106**	**20,334**	**19,309**	**16,484**	**16,474**	**17,131**	**16,897**
For human food	8,006	12,662	13,242	13,821	13,714	13,584	13,625	13,739	14,175
For industrial use	3,351	3,687	2,864	6,513	5,595	2,900	2,848	3,392	2,722
Domestic catch	**6,482**	**9,404**	**9,637**	**10,467**	**10,461**	**9,788**	**9,565**	**9,845**	**9,194**
For human food	3,654	7,041	7,618	8,214	7,936	7,667	7,476	7,248	7,174
For industrial use	2,828	2,363	2,019	2,253	2,525	2,121	2,090	2,597	2,020
Imports [1]	**4,875**	**6,945**	**6,469**	**9,867**	**8,848**	**6,696**	**6,909**	**7,286**	**7,703**
For human food	4,352	5,621	5,624	5,607	5,778	5,917	6,150	6,491	7,001
For industrial use [2]	523	1,324	845	4,260	3,070	779	759	795	702
Disposition of domestic catch	**6,482**	**9,404**	**9,637**	**10,467**	**10,461**	**9,788**	**9,565**	**9,846**	**9,194**
Fresh and frozen	2,621	6,501	7,288	7,744	7,475	7,099	7,054	6,877	6,870
Canned	1,161	751	543	649	622	769	678	648	516
Cured	96	126	100	115	95	90	93	108	129
Reduced to meal, oil, etc.	2,604	2,026	1,696	1,959	2,269	1,830	1,740	2,213	1,679

[1] Excludes imports of edible fishery products consumed in Puerto Rico; includes landings of tuna caught by foreign vessels in American Samoa.   [2] Fish meal and sea herring.

Figure 10.2. U.S. Census Bureau, *Statistical Abstract of the United States: 2000* (120th edition) Washington, DC, 2000, p. 692. Table data source, U.S. National Oceanic and Atmospheric Administration, National Marine Fisheries Service, Fisheries Statistics and Economics Division, Silver Spring, MD.

the United States, which provides information on the amount of fish that has been caught in or imported to the United States between 1980 and 1998. The explanatory material tells that all the numbers listed in the table should be multiplied by one million. Therefore, the total domestic catch of fish in 1980 was 6,482,000,000 pounds and the domestic catch in 1998 was 9,194,000,000 pounds.

Legal Information

Legal information can be difficult to track down since the legal system is constantly evolving. In scientific research, however, it may be very important to follow the progress of legislation relating to a particular area of research. Genetic engineering, nuclear energy, and environmental protection are examples of areas that are often under consideration for legislative change.

In the United States, the laws of the land are published in the *United States Code.* This document is fully revised only every six years. The *Statutes at Large* may be used to track legislative changes between editions of the *U.S. Code.* This resource contains a chronological compilation of slip laws, the first publication of legislative bills. Slip laws are later codified and entered into the *United States Code.* The various agencies of the Executive Branch enact rules and *regulations* to enforce the laws passed by Congress. These rules and regulations are cumulated annually into the *Code of Federal Regulations (CFR).* The *Federal Register* is published daily, Monday through Friday, to track proposed and recently enacted regulations of the executive agencies. The *U.S. Code,* the *Code of Federal Regulations,* the *Federal Register,* and the text of the slip laws are all available on the Web through *GPO Access.*

Another useful resource for locating information on federal legislation is *Thomas,* a Web site maintained by the Library of Congress. *Thomas* is located at http://thomas.loc.gov/. At this site you can find the summary and status of bills, you can search for Public Laws and you can search committee reports by keyword. If you need to know how your own legislator voted on a key issue, you can check out the roll call link that is available from this site.

Obviously, federal, state, and international law research is extremely complex and cannot be adequately addressed within a few paragraphs. Legal research is an area where it is tremendously helpful to seek assistance from a librarian who specializes in legal or government regulations. He or she can help you to determine whether you are accessing the latest version of a law or regulation and how to track changes that may be made to the law in the future.

State laws and regulations generally follow similar patterns of publication and codification. Since state and local laws vary from one location to another, a legal or government documents librarian is also your best resource for help in locating these materials. Access to state legal information may be available from the government documents section of your library, or on the state's official Web site. A good source for state legal information, if your library subscribes, is *Lexis-Nexis Academic Universe.*

Patents

Patents are another important source of scientific and technical information published by the United States government through the *Patent and Trademark Office (PTO)*. Patents grant an inventor the right to prevent others from manufacturing or selling an invention for a certain number of years. In exchange, the inventor must publish the details of the process or invention, with drawings. A patent must include one or more *"claims,"* or descriptions of what makes this particular invention different from all others. A patent must also include an analysis of "prior art," which is a summary of the technology leading up to the proposed invention. Patents may be issued for a device, process, or "composition of matter" (utility patents), for the appearance of an object (design patents), or for vegetatively reproduced plants (plant patents). Much of this technology is never published in journals or presented at conferences. With over six million patents issued since 1790, U.S. patents are a tremendous store of technical information.

The bibliographic information, one drawing, and one claim of all newly issued patents are announced in the *Official Gazette of the United States Patent and Trademark Office: Patents* (SuDocs number C 21.5:) issued every Tuesday by the PTO and available at most GPO depository libraries. Electronic searching of the patent literature is conducted through a database called *CASSIS* issued by the PTO to its depository libraries or for a fee through commercial databases. The Patent and Trademark Depository Library (PTDL) program is separate from the GPO depository library program described earlier in this chapter. Until recently, you had to visit one of 61 patent depository libraries in order to search patent literature; now everything you need is available on the Internet through the PTO's Web site, http://www.uspto.gov. Patents issued since 1976 can be searched using such fields as title words, inventor name, class and subclass, and, of course, patent number. Patents issued between 1790 and 1975 can be searched only by patent number or by class and subclass. All patents since 1790 have images of the actual patent, including drawings, on the Web.

While the inventor is required to make the details of an invention public, nothing says that she has to make it easy for you to find it. Titles of patents are usually very vague and generic. If you are looking for the proverbial better mousetrap, you will have to look for a "Rodent trap with diverter" (Pat. No. 6,050,024), a "Pest control system" (Pat. No. 6,266,917), or simply a "Trap" (Pat. No. 5,515,642). The PTO therefore uses a hierarchical system of classification, the *Manual of Classification* (SuDocs number C 21.12:), which is available from GPO depository libraries and on-line at the PTO Web site. Our mousetraps are listed in class 43 ("Fishing, Trapping, and Vermin Destroying") and subclass 81 ("swinging striker"), written "43/81," along with other classes and subclasses describing other aspects of the inventions. Other tools for searching patents are the *Index to the U.S. Patent Classification* (SuDocs class C 21.12/2:) and the class definitions.

Patents are sometimes confused with trademarks and copyrights. Each of these is designed to protect intellectual property, but each protects a slightly different aspect. *Patents* protect inventions, *trademarks* protect the identity of a product or service, *copyrights* protect the form and content of literary or other creative works, and *trade secrets* protect formulae. Trademarks are also registered with the Patent and Trademark Office, copyrights are registered through the Library of Congress, and trade secrets, of course, are not registered anywhere. Some inventions may qualify for multiple registrations. Software, for instance, may be both patented and copyrighted. Trademarks, though important for business, are not a source of scientific or technical information.

Locating Government Agencies and Officials

At times, you may need to find information about a particular government agency or locate the names of officials in one of these agencies.
The first place that many researchers turn for this kind of information is
the *United States Government Manual*. This resource is published annually and provides basic information about agencies and offices of the
executive, legislative, and judicial branches of the United States government. Independent government agencies such as the National Science Foundation and the Federal Emergency Management Agency are
also included. The latest editions of the *U.S. Government Manual* are
available through *GPO Access*. Since the *U.S. Government Manual* is
published on an annual basis, it will not list any name changes that
have occurred since the latest edition was published. This can be a particular problem during the first year of a presidential term since government officials tend to change with the new administration. The *U.S.
Government Manual* may still be used, however, to locate the phone
number and Web address of the agency, which can then be used to find
out the latest information about the agency and its officials.

State and local officials may be more difficult to locate but, thanks
to the Internet, research in this area is usually not as difficult as it once
was. Many states have good Web sites that provide easy access to state
agencies, officials, and information on state legislation. Use your favorite Internet search engine and try searching for the name of the state
and the word "government." Once you locate a specific site, be sure to
use the Internet evaluation techniques discussed in chapters 9 and 11.
Many sites may provide information about state governments, but you
should try to locate the official site for the state. The Web address for
an official state home page often follows the format http://www.
state.stateabbreviation.us where the state abbreviation is usually a two
letter code representing the state. For example, the official home page
for the state of Indiana is: http://www.state.in.us/ and the official home
page for the state of Georgia is: http://www.state.ga.us/. A final word of
caution when looking up names of state government officials: when
using an Internet search on house or senate representatives, it is easy to
locate the *federal* representative when you are actually seeking the *state*
representative. A close examination of the Web site should give you an
indication of whether you have retrieved the federal or state official.

Summary

An essential part of any scientific literature review includes a search of the government documents. Depository libraries provide easy access to *federal documents,* and government documents are also available through interlibrary loan or may be purchased from the Government Printing Office or from other government disseminating agencies such as NTIS. Federal documents are arranged by issuing agency and are normally classified and shelved using a classification system called the Superintendent of Documents Classification System. Primary indexing to documents is contained in the *Monthly Catalog of United States Government Publications.* Other resources, such as *GPO Access, NTIS, GrayLIT Network,* and *WorldCat,* provide additional access to government reports.

The United States government is also a primary source for a great deal of statistical information. A wealth of information is available in the *Statistical Abstract of the United States.* When using statistical tables, it is important to check the explanatory material that accompanies the table. This section will provide any footnotes and comments on the units of measurement used as well as a citation to the source of the original data.

United States law is contained in the *U.S. Code.* Rules and regulations on how to enforce the laws are available in the *Code of Federal Regulations* and in the *Federal Register.* Since the law is continually being evaluated and updated, it is necessary to use great care when doing legal research so that the most current information is located. Consulting a legal reference librarian or a government documents expert is highly recommended for this type of research.

Patents are another source of scientific and technical information available through the government. Patents can now be searched effectively on-line through the Patent Office's Web site.

Chapter 11

What Have I Found?
Evaluating Information

Does this sound familiar? "I found it in the library so it must be reliable." "I found tons of great stuff on the Internet!" How do you know that the information you plan to use, in academic research, in business ventures, or in personal decisions, is really accurate and reliable? If you have studied or worked in a field for a period of time, you may recognize certain individuals or organizations as reliable sources. The explosive growth in publishing of both print and electronic material makes it increasingly difficult to recognize all sources on sight. Even if the material is reliable, how do you know if it is appropriate for your intended use? Is it "scholarly" enough? Is it written for college students, for a knowledgeable working adult, or for a high school student?

A library collection specialist attempts to acquire materials that meet the needs of the users. Although evaluation procedures are in place and provide some control over quality, realistically even the library cannot verify the accuracy of all the materials it collects. The Internet defies many traditional quality control measures. Nearly anyone can create and maintain a Web site on any topic that strikes his fancy. And they do! A "simple" search of the Internet is an oxymoron.

This leaves the burden of assessment on the information user. We will look at some criteria for evaluating information and at some tools that can help determine the value of that information.

Learning about the Author

Gathering some details about the author can be one of the most important steps in evaluation. What is the educational background of the author? What kind of professional experience has he or she had? Is he or she associated with any relevant professional organizations? Does the author's background make him or her a reliable source of information in the area of study?

A variety of biographical locating tools can be used to find information about an author. *Biography and Genealogy Master Index (BGMI)* indexes many biographical directories and dictionaries. *BGMI* itself provides little, if any, actual biographical information. Its purpose is to lead you to biographical sources, such as *Who's Who*, *Contemporary Authors* and *Current Biography*, which will provide you with more detailed information. *BGMI* is easy to use. Simply look up a person's name and, assuming the person is famous enough, you will find a list of titles and editions of biographical resources that contain information on that person. Be sure to pay attention to the edition or year cited along with the title of the resource. Many biographical resources are issued on an annual basis so, while your person may appear in the 1987 edition of that resource, he or she may not be listed in the 1994 edition. Your next step is to go to your library catalog to find out if your library owns the particular titles and editions listed in the *BGMI* entry.

Biographical resources vary in the amount of information they provide about an individual. Biographical dictionaries contain a brief mention of a person and provide a few essential facts. The *Who's Who* series give a bit more information about the person's educational and professional career. *Who's Who* or similar dictionary-like biographical resources will normally indicate where and when the person received his education. It will also tell if he or she belongs to professional organizations and will list awards and honors that he may have received, such as a Nobel Prize or admission to the National Academy of Sciences. Brief biographies may also provide the author's occupational history—listing places of employment and the length of time at each location. Subject-specialized biographical resources such as *American Men and Women of Science* or *Who's Who in Engineering* are also available. These resources limit their coverage to people who have reached some level of recognition among their peers.

Current Biography is a well-known biographical resource that provides a longer article about an individual and his or her life's work, and often includes a brief bibliography of additional resources. To find book-length resources about a particular person, refer to your library catalog or to *WorldCat* and perform a subject search on the person's name.

What if you can't find out *anything* about a particular author? In some cases, you may be unable to locate information about your author in a standard biographical source. This does not mean that the information in the article is unreliable. An author may be new to the field or just beginning her career. In other cases, the author may be an irregular but competent contributor or a knowledgeable amateur. Everyone has to get started sometime and does not have to be a household name to have much to contribute. Determining the author's status may be, in itself, of value to the information seeker. Besides, there are other ways to evaluate materials.

Sometimes background or educational information may not be sufficient to evaluate an author. In this instance, investigating the quantity or quality of the author's work is helpful. If you found an interesting article by browsing through a periodical, select one or two subject-appropriate periodical indexes and perform an author search to see if he has published anything else. Compare the subject content of some of these articles. The author may turn out to be a regular columnist for a publication. Perhaps the author is a researcher who publishes on an irregular basis.

Corporate Authors

Don't be put off by "corporate authors," i.e., organizations that produce a document but do not identify a specific individual responsible for the material. In this case, you will want to evaluate the organization rather than the individual. Consult appropriate directories, periodical and newspaper articles, and reports to obtain background information on the mission and objectives of the organization.

Evaluating the Publication

The actual publications can help to evaluate the quality of the information. There are many different types of periodicals; some target the general population, others appeal to subject specialists. Some are *trade magazines* that may place more emphasis on news and current trends in a particular field. Others may stress scholarly research.

There are several levels of scholarly journals. Many journals are considered to be appropriate for scholarly research if they are formatted in a scholarly manner, that is, the articles describe sound research techniques and present a bibliography. The most scholarly periodicals, though, are the peer-reviewed journals. These publications, sometimes referred to as juried or refereed journals, require manuscripts to pass a stringent peer review process prior to publication. The manuscripts are sent out to experts in the field who read them and decide if the information included is valid and worthy of publication. The most stringent form of peer-review is the *double blind review*. In this case, the name of the author of the article is not given to the reviewer and the reviewer's identity is unknown to the author. This helps to eliminate some of the biases that can develop for or against a particular author's work. In chapter 8, we described two research tools: *Ulrich's International Periodical Directory* and *Magazines for Libraries*. These tools can be used to locate the scope and content of many magazines and journals and will indicate if a journal is peer-reviewed.

Book publishers, like journal publishers, may develop a reputation for quality. Some publishers, such as Wiley or Blackwell Scientific, specialize in scientific publishing. University presses often publish scholarly materials. On the other hand, vanity publishers may be less critical of something if the author is willing to pay for it. A reference librarian can normally help you to determine the credibility of a publisher.

Citation Indexes

It is one thing to get published and another to be respected for one's work. This may be evident in how often other authors refer to an individual's work or how well it is received. There are ways to measure this activity.

In chapter 8, we discussed how to locate periodical articles using citation indexes. Citation indexes also offer a specialized search feature called a cited reference search, which allows you to search the bibliographies of published research. You can search this portion of the database by author's name, by the title of the publication where the article appeared, or by the date of publication. In figure 11.1, we have entered a search for an article by D. Boran that appeared in the journal *Marine Chemistry* in 1983. Remember, in *Web of Science* you do not search by the full name of the author. We enter the last name of the author, "boran" and the author's first initial, "d." Since we do not know the author's middle initial, we have entered an asterisk as a wildcard. The "cited work" box should contain the name of the publication that contains the article in question. A search box example is available to illustrate that the publication title should be entered as an abbreviation—but not just any abbreviation. It must be entered according to the list that the Institute for Scientific Information (ISI) has compiled. A list of authorized abbreviations is available via a hot link on the search screen. According to this list, the journal *Marine Chemistry* is abbreviated as "mar chem." Finally, enter the date of publication in the "cited year" box.

Figure 11.2 illustrates the full record that is obtained from the search on Boran and *Marine Chemistry*. The full record shows that the actual title of the article is "The structure of marine fulvic and humic acids." We see that the article was written by multiple authors, Harvey being the primary, or first author, and Boran and others being the secondary authors. The ability of the user to search in *Web of Science* and other scientific databases for both primary and secondary authors is very valuable. When recommending a particular piece of research, it is not always easy to accurately recall the order of authorship or the names of all of the authors. The flexibility of the author search expands the researcher's potential for locating relevant items.

Returning to the illustration in figure 11.2, we can also note that the article appeared in volume 12, issue number 2-3, of *Marine Chemistry*. The issue was published in 1983 and the article appears on pages 119-132. The article is written in English. This information is standard and can be located in several other specialized periodical indexes. What makes *Web of Science* particularly interesting is the additional information that can be located under "Cited References" and "Times Cited."

Figure 11.3 illustrates the first section of information that is retrieved when you click on the "Cited References" link. Listed at this link are the citation details for the twenty-eight references that make up the bibliography of the *Marine Chemistry* article. This feature provides a historical path to information that was relevant to the research performed in the writing of this article.

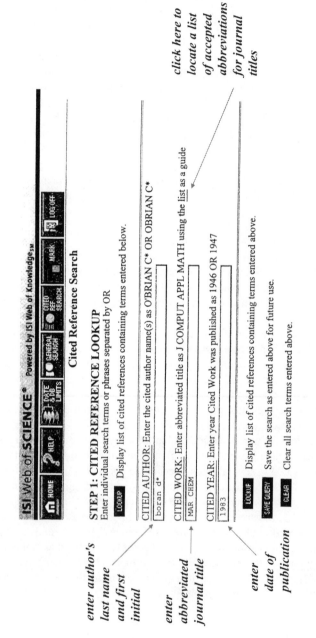

Figure 11.1. *Web of Science* **cited reference search screen.**
Copyright © 2001 Institute for Scientific Information.® Reprinted with permission.

Figure 11.2. Main entry from a *Web of Science* **cited references search.**
Copyright © 2001 Institute for Scientific Information.® Reprinted with permission.

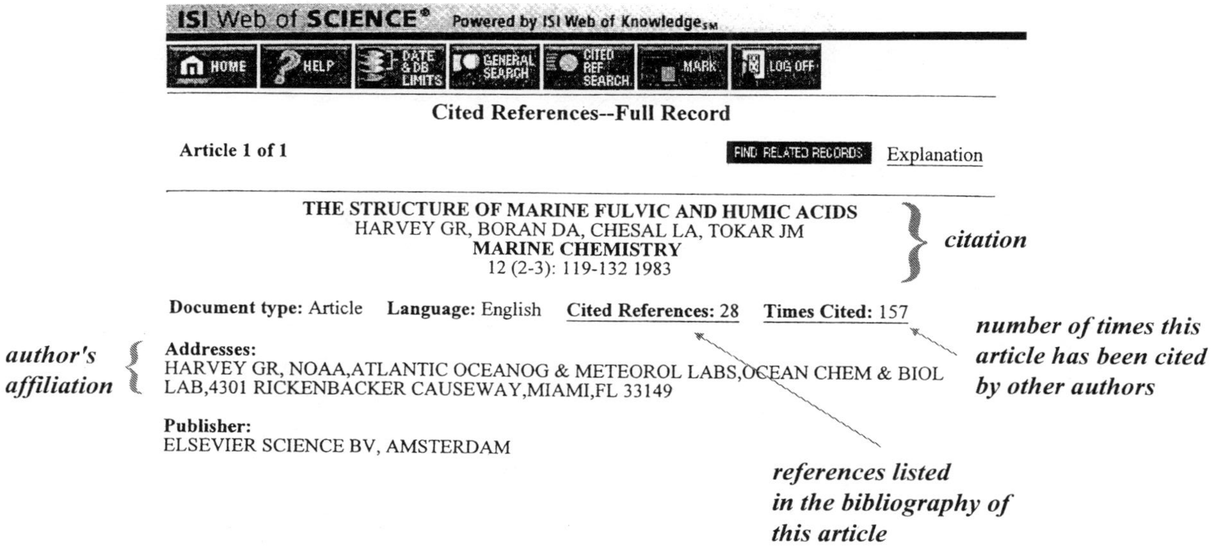

Looking back at the screen for the full record, we now click on the "Times Cited" link. The first few citations from this search are shown in figure 11.4. The "Times Cited" link shows a citation for every resource that was included in the ISI databases that has used the Harvey et al. article in its own bibliography. This provides a method for locating more recent research that should be related in some way to the original article. Additionally, the number of times that an article has been cited by other authors can be used as a method of establishing credibility for the original research. It is, of course, possible that the later articles are citing the original article because they feel that it is incorrect. That doesn't happen often, however, and it is usually assumed that a large number of citations by others indicates that the research is well thought of. Also, some authors have a tendency to cite themselves in later articles, which inflates the number of times that their original article is cited. It is not unusual for an author to cite his own work occasionally—after all, his future endeavors are usually built on what has already been accomplished, but some authors can get carried away with the idea. In this case, the article has been cited an impressive 157 times by others and it continues to be cited as evidenced by the very recent entries. Obviously, this article includes key research for the field.

Book Reviews

Book reviews tell something about the quality of information published in book format. Reviews may vary in length, some may be more scholarly than others, or may target a specific audience. The reviewer's credentials also vary. These factors depend greatly on the type of publication offering the review. For example, *Library Journal* publishes short reviews, written by librarians, targeting library collection managers who are purchasing materials for various kinds of libraries. *Nature*, a scientific publication, reviews a wide variety of books that are of general interest to the science community. *Earth Science Reviews* offers reviews of books in a specialized area of science. Two primary tools for locating book reviews are *Book Review Index* and *Book Review Digest*. While *Book Review Index* is more comprehensive, the *Digest* contains

Cited References

THE STRUCTURE OF MARINE FULVIC AND HUMIC ACIDS
HARVEY GR, BORAN DA, CHESAL LA, et al.
MARINE CHEMISTRY
12 (2-3): 119-132 1983

FIND RELATED RECORDS   Explanation } *original citation*

Clear the checkbox to the left of an item if you do not want to search for articles that cite the item when looking at Related Records.

Cited Author	Cited Work	Volume	Page	Year
☑ ACKMAN RG	J FISH RES BOARD CAN	25	1603	1968
☑ BAKER N	J LIPID RES	7	341	1966
☑ BAKER N	J LIPID RES	7	349	1966
☑ BOON JJ	GEOCHIM COSMOCHIM AC	44	131	1980
☑ FUOSS RM	J POLYM SCI	3	602	1948
☑ GAGOSIAN RB	MAR CHEM	5	605	1977
☑ GSCHWEND PM	LIMNOL OCEANOGR	25	1044	1980
☑ HARVEY G	MAR CHEM	8	327	1980
☑ HATCHER PG	NATURE	285	560	1980
☑ HATCHER PG	ORG GEOCHEM	2	77	1980
☑ KERR RA	DEEP-SEA RES	22	107	1975
☑ KIRSCHENBAUER HG	FATS OILS		CH4	1960

Figure 11.3. *Web of Science* cited references link showing citations to articles that were included in the bibliography of the 1983 *Marine Chemistry* article.

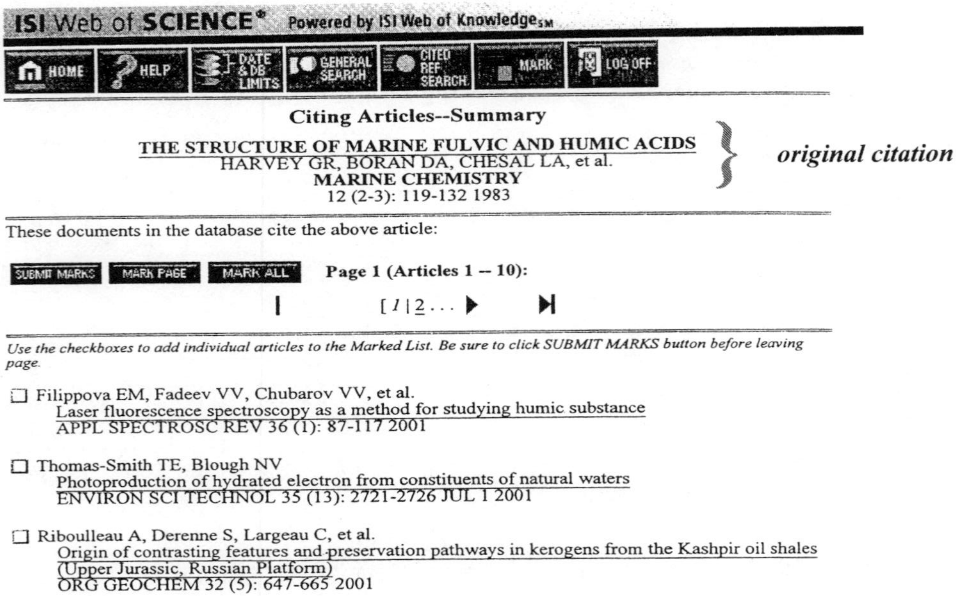

Figure 11.4. *Web of Science* "Times Cited" link showing citations for recent articles that have cited the 1983 *Marine Chemistry* article in their bibliographies.
Copyright © 2001 Institute for Scientific Information.® Reprinted with permission.

excerpts of reviews. Many other indexes and databases include book reviews as well. The H.W. Wilson Publishing Company includes book reviews in many of their databases. Reviews can also be located in *Web of Science*.

When searching for a review, first determine when the book was published. Begin your search with an index that includes the year of publication and several years following that date. The time involved in the reviewing and indexing process means that some reviews may not appear in an index until several years after the book has been published. Most indexes to book reviews allow you to search by either the author or the title of the book. As with all bibliographic indexes, once you find the citation to the review, you will need to go to your library catalog to determine if the complete text of the review is available in your library.

Other Considerations

Quality and style of writing should be considered by looking for the presence of bias, the level of vocabulary used, the proper use of grammar, and the intended audience. Accuracy is, of course, critical. You may find it necessary to confirm some details or facts in more than one source rather than assume any one source is correct. The author must effectively organize and communicate the information to you, the intended audience. With careful reading, you should be able to understand what the author is trying to say. The correct use of grammar and an appropriate level of vocabulary are indicative of carefully prepared material. While bias may be in the eye of the beholder, if the material purports to represent an evenhanded view of an issue or topic, obvious bias should be avoided.

Other useful features are bibliographies or lists of the references used in the preparation of the item. These bibliographies not only lead the information seeker to additional materials, but also may help you in verifying facts. Timeliness of the information provided in the text and of the references used should be noted, even on subjects of a historical nature. Does the author show that he or she is aware of current research on his or her subject? For example, although the theory of relativity was first introduced many years ago, new information on this topic continues to be published. New research and insights on historical events may be as important as the original material published on a subject.

Evaluating Internet Resources

The same criteria that have already been discussed can and should be applied to information found on the open, freely accessible material on the Internet. Remember, proprietary, fee-based on-line resources have passed several levels of review, first by the publisher and then by the purchaser of the data. As with any resource, you shouldn't blindly accept the information that you find on the open Internet. Because of the unique nature of the Internet, especially the highly graphical World Wide Web, some modification of criteria and a few extra tidbits should be discussed.

Author

The Internet author has the same obligation to organize and present his or her information in a clear and understandable way, as does the author who appears in print. Examine the credentials of the author, producer, or organization responsible for the site. You might also pick up some ideas from the Web address. As mentioned in chapter 9, the Web domain can give you an indication of the type of organization that sponsored the author of the Web page. For example, addresses ending in .gov, .com, and .edu, represent governmental, commercial, and educational organizations, respectively.

Content

A good Internet site should state or imply its purpose or mission. Critically review the quality of writing, including grammar, vocabulary, and accuracy of the information provided. A Web site can be organized in an almost infinite number of ways. How well the material is organized is one criterion that can be used to evaluate the site. The searcher usually proceeds from one item of information to another with a click of a mouse on hot links, which may be highlighted titles, names, or labels. If the content of a link is not self-evident from its title, an annotation or description of the site is a welcome feature. You should ask if the links are applicable to the purpose of the site. Does the site present a selective group of links or does it list many links in an attempt to be comprehensive? In either case, how well has it achieved its goals?

Reviews

Many Web sites are reviewed by outside reviewers just like books and articles. Look for these reviews in journals or on the Web. References to Web reviews may be advertised with the appearance of icons or graphics on the page. There are many different types of organizations that review Internet sites, and awards are frequently listed on the home page of the recipient.

Timeliness

Another problem with the Web is that it can be difficult to determine how recently the material was published. Some Web sites include a "date last updated" section on the page. If that is not available, you may be able to determine the timeliness of the page by looking at the information that is provided. If dates are mentioned in the text, are they recent? Are current events listed still "current"? Are the links to other sites still active?

Bias

Evaluating a site for bias is particularly important when examining an Internet resource. Because there normally isn't an editorial review, bias can be easily incorporated into a Web site. If a site is a commercial site that is promoting its own product, you can expect it to be somewhat biased towards its own company. On the other hand, an organization can put forth a subtle agenda that may not be easy to spot. Topics such as cloning or stem cell research are controversial, and a Web site may have been created with an unstated intent to sway the viewer's opinion on the topic. Think about the kind of information that is being presented and determine whether both sides of an issue are being treated fairly.

Structure

Because of the highly graphical nature of the Web, a site can be a little overwhelming with fancy moving, flashing, popping images that may have nothing to do with content or function. Look at the design of the site. Does the use of graphics flatter the page and serve a function, or are they an unpleasant distraction? Are the functions of the icons clear? If there is a search engine associated with the site, is it easy to use and efficient?

Evaluation Isn't Always Easy!

Everything we have discussed in this chapter may be used to evaluate materials you intend to use for your research; even with all of this information, however, it may still be tricky to assess the value of a resource. Examine the Web site illustrated in figure 11.5. This is the home page for DHMO.org. The site is well organized and seems to be quite informative. It tells us that DHMO.org is the Web site for the Dihydrogen Monoxide Organization. The site is authored by the Dihydrogen Monoxide Research Division, which is located in Newark, Delaware. On the home page we learn that the "goal of this site is to provide an unbiased data clearinghouse and a forum for public discussion." Links are included that specify the dangers associated with DHMO. DHMO.org has useful links to additional, well-known sites such as the Environmental Protection Agency, the National Institute of Health, and the Centers for Disease Control and Prevention. It also contains a useful link to a frequently asked questions *(FAQ)* page, which is illustrated in figure 11.6. The links are active, indicating that someone is continuing to maintain the site, and the site was last updated in 2001 so it remains current.

Closer examination of this site will reveal that it is an extremely clever satire. The site appears authoritative, well organized, and up-to-date. The information given on the site is undoubtedly true; however, the chemical that the organization is concerned about is water!

Dihydrogen Monoxide - DHMO Homepage

Translations ▼

Translations ▼

 United States
Environmental
Assessement
Center

Dihydrogen Monoxide
Research Division

Support the cause! Visit the DHMO.org Store

DHMO Special Reports

- Dihydrogen Monoxide FAQ
- Environmental Impact of DHMO
- Dihydrogen Monoxide and Cancer
- DHMO Surveys & Research

- DHMO in the Dairy Industry
- DHMO Conspiracy
- Editorial: Truth about DHMO
- Fake Email SPAM Alert
- Linking to DHMO.org

- Alerts & Advisories **NEW**
Sign-up to receive periodic safety bulletins from DHMO.org.

| you@domain.com | OK |

Visit the **DHMO.org STORE**   T-shirts, HazMat Vials, Pamphlets and more!   **CLICK HERE**

WELCOME

Welcome to the web site for the Dihydrogen Monoxide Research Division (DMRD), currently located in Newark, Delaware. The controversy surrounding dihydrogen monoxide has never been more widely debated, and the goal of this site is to provide an unbiased data clearinghouse and a forum for public discussion.

Explore our many Special Reports, including the DHMO FAQ, a definitive primer on the subject, plus reports on the environment, cancer, current research, and an insider exposé about the use of DHMO in the dairy industry.

The success of this site depends on you, the citizen concerned about Dihydrogen Monoxide. We welcome your comments and suggestions.

Send us your feedback!

Due to the high volume of email we receive, we may not be able to reply to every letter. However, we do read them all.

DHMO Related Info:

Media Press coverage

National Consumer Coalition Against DHMO

Environmental Protection Agency

NIH National Toxicology Program

Centers for Disease Control & Prevention

American Cancer Society

Sierra Club

Greenpeace

Send Email to Your Representative

Figure 11.5. Dihydrogen Monoxide: DHMO Homepage http://www.dhmo.org.
Copyright © 1997-2001 Tom Way. Reprinted with permission.

Dihydrogen Monoxide FAQ

Frequently Asked Questions About Dihydrogen Monoxide (DHMO)

What is Dihydrogen Monoxide?

Dihydrogen Monoxide (DHMO) is a colorless and odorless chemical compound, also referred to by some as Dihydrogen Oxide, Hydrogen Hydroxide, Hydronium Hydroxide, or simply Hydric acid. Its basis is the unstable radical Hydroxide, the components of which are found in a number of caustic, explosive and poisonous compounds such as Sulfuric Acid, Nitroglycerine and Ethyl Alcohol.

For more detailed information, including precautions, disposal procedures and storage requirements, refer to the Material Safety Data Sheet (MSDS) for Dihydrogen Monoxide.

Should I be concerned about Dihydrogen Monoxide?

Yes, you should be concerned about DHMO! Although the U.S. Government and the Centers for Disease Control (CDC) do not classify Dihydrogen Monoxide as a toxic or carcinogenic substance (as it does with better known chemicals such as hydrochloric acid and saccharine), DHMO is a constituent of many known toxic substances, diseases and disease-causing agents, environmental hazards and can even be lethal to humans in quantities as small as a thimbleful.

Research conducted by award-winning U.S. scientist Nathan Zohner concluded that roughly 86 percent of the population supports a ban on dihydrogen monoxide. Although his results are

Figure 11.6. Dihydrogen Monoxide: DHMO Facts
http://www.dhmo.org/facts.htm.

Summary

Does this mean that every single piece of paper, book, magazine, or whatever should be subjected to this process? Despite all that has been outlined previously, the answer is no. In many cases, you may rely on your own good judgment. Certainly, information that appears consistently from one source to the next is probably reliable; and, if you may recall, we mentioned that in your field of research there might be individuals, organizations, or publications that are known to be reliable and accurate.

Evaluation becomes especially important when you venture into areas with which you are not familiar, or perhaps where new evidence is being presented, where information may be provocative, where bias is suspected, or simply, whenever it is *imperative* that all information must be accurate and defensible.

Tools for evaluating resources:
- Biographical information
- Citation indexes
- Book reviews

Things to consider when evaluating information:
- accuracy
- publisher
- peer-review
- credentials of the author
- timeliness of the material presented
- quality of writing
- intended audience
- lack of apparent bias

The sample research strategy included in appendix 2 will provide additional examples on how to evaluate information.

Chapter 12

Where Do I Go from Here?

It is our hope that the concepts illustrated in this book will be just as useful to you after graduation as they are now. The same theories that are used to research term paper topics can also be used to locate financial information, to find ideas for future employment, to research health concerns, or to locate good spots for your next vacation!

Identifying Graduate Schools

A key decision facing many upperclassmen in the field of science is deciding on an appropriate graduate school. In addition to suggestions from faculty members and scientists in the field, the library has many resources that can be helpful for finalizing your choice. The *Peterson's* series on graduate education is a particularly useful resource for finding details on research departments at specific institutions of higher learning. Three volumes are of special interest to future scientists: *Peterson's Graduate Programs in the Physical Sciences, Mathematics, Agricultural Sciences, the Environment & Natural Resources; Peterson's Graduate Programs in Engineering & Applied Sciences;* and *Peterson's Graduate Programs in the Biological Sciences.* Organized by subject, the entries in each volume list specific degrees awarded (M.S., Ph.D., etc.), entrance requirements, and basic tuition costs, as well as application addresses and deadlines. Selected institutions have more detailed entries that include a listing of faculty and their research projects. The prefaces to the *Peterson's* books also contain helpful tips on the graduate application process and financial aid possibilities.

Once you have selected several institutions that look promising, you will want to get more familiar with the research goals of the university and its faculty. The university home page is an excellent resource for current information about the school. The World Wide Web address for many universities is available in numerous directories, including the *Peterson's* guides. The Web addresses may also be located by performing an open Internet search. From the home page, locate the link to the department or college in which you are interested. Many universities provide links from departmental pages to faculty *curriculum vitae* and research information.

Another method for locating information on research at a particular institution is to use one of the electronic databases. A number of databases allow you to search by the name of the institution, either alone or in combination with a topic keyword or an author's name. Suppose you want to know more about the type of research being conducted in the field of immunotherapy at Yale University and, in doing so, identify the names of the faculty researchers in that field. A search in *Web of Science, Medline,* or *SciFinder Scholar* may yield articles that will provide you with that information.

Dissertation Abstracts indexes every doctoral dissertation published by an accredited institution in the United States, as well as some Master's theses and foreign language theses. Searching by institution along with appropriate keywords may yield a helpful set of records that define the kind of research being conducted at that institution and identify the faculty who direct and advise others in that work. Keep in mind that older records may not identify advisors, and indexing with subject headings in this database is minimal.

Once you have selected schools, you will want to create an impressive application. Many handbooks and manuals have been written that provide tips on gaining admission to graduate schools and on financing your higher education. There are an extensive number of books that deal with financial aid, grants, and fellowships. The titles may vary from one library to the next, but most will be found under the Library of Congress subject heading:

universities and colleges—united states—graduate work

Perform a subject search on your library on-line public access catalog (OPAC) to locate these materials.

Employment Opportunities

There are many employment opportunities for students who major in the sciences or engineering, including positions at academic institutions, government entities, and industries. In the academic world, the scientist often researches and teaches. To conduct research, you must have not only a solid understanding of fundamental principles and processes in your field but must also keep abreast of the advancements made by others. These same qualifications enhance your ability to teach the subject effectively to others. The educational background necessary for employment in academia may vary considerably with the type of position, the responsibilities, and the level of skill required. For a laboratory staff line, a bachelor's degree might be a minimum qualification; senior staff and faculty positions would require a master's or doctoral degree.

Federal, state, and local governments hire scientists and engineers in many capacities. Government scientists may perform basic and applied research, which may be used to help establish policy guidelines, standards, or regulations. Often, these scientists are also involved in educational activities for the nation or local communities.

Private industries employ many scientists and engineers to conduct basic and applied research to develop new or to improve existing products, equipment, or services that the company can market to consumers. As is the case with academia, government and industrial scientists may often have advanced degrees. In all cases, salary is often dependent on the level of education achieved and number of years experience in the profession.

Grant-Writing Tools

A scientist's salary is often supported by the organization that employs him or her, but funding for the various research projects that he or she wishes to pursue may come from other sources. Graduate students, post-docs, and advanced undergraduates may also consider looking for their own grant funding to pursue a particular research project. National, state, and local government agencies fund a great deal of research. Other sources of financial support may come from industry or private foundations, or a partnership of one or more of these organizations.

The National Science Foundation (NSF) reports that it provides approximately 20 percent of all federal support to academic institutions.[1] The NSF Web site, http://www.nsf.gov, contains current information about funding opportunities, instructions on preparing and submitting a proposal, deadlines, and related details. Other government agencies such as the National Aeronautics and Space Administration (NASA) and the National Institutes of Health (NIH) maintain similar instructions on their Web sites. Notices of funding availability (NOFA), which appear in the *Federal Register*, are also available at http://ocd.usda.gov/nofa.htm. State and local governments may have funds to award. Many receive funds from the federal government that are, in turn, handed down for studies that are regionally or locally based. Check state and local government Web sites for leads or contact agencies directly.

Many scientific societies award their own grants and also announce the availability of funding from other sources. For example, the American Association of Petroleum Geologists (AAPG) funds graduate student research in the geosciences.

Your library collection may include helpful directories such as *The Foundation Directory* and the *Annual Register of Grant Support*, which facilitate the identification of many possible funding sources including public and private organizations and businesses. These directories are well organized, providing subject headings, names, and geographical indexes.

Submitting a Manuscript or Article

Regardless of the employer, scientists are often expected, if not required, to publish their work in books, scholarly or professional journals, or report documents. Many publishers specialize in certain subject areas and have guidelines that must be followed if one expects to see his or her efforts in print. It is preferable to have a contract with a publisher before you write and finish your entire book. While you have the expertise in a certain field, a good publisher has the knowledge and experience to market your book successfully.

Appropriate publishers may be identified by using a trade publishing resource such as *Literary Market Place, Books in Print*, or by noting the publishers of other scientific books in your field. Guidelines vary, but usually a letter of inquiry to a publisher will also include some sample chapters, illustrations, and some biographical information about the author. These requirements can usually be found on the publisher's Web site. Based on this information, the publisher will decide whether your book idea and writing style are consistent with its mission. First-time authors should not be discouraged if rejected by the first publisher. Often, it is a matter of finding a good match between the book content and the publisher's subject coverage.

Publishing in scholarly or professional journals differs from publishing in books in that the scientist hopes to present his or her current research findings in a timely way. Once the research project is completed, the scientist prepares a manuscript for publication. Finding a suitable journal depends on the subject area, type of research, and outcomes. Most journals publish all or part of their requirements for manuscript submission; identify fees or charges if applicable; specify reprint availability and the number of free reprints; and state the all-important copyright information of which all authors need to be aware. Potential authors will often be referred to a Web site for complete instructions, including manuscript format and the journal style guide. Chapter 2 provides information on style manuals and how to cite resources properly. The journal usually identifies whether or not it subjects all submissions to a peer-review process. Depending on the publication frequency of the paper journal, the author may still find that it takes a considerable amount of time (one, two, or more years) for the article to appear. There are journals that attempt to expedite the process, not necessarily by dropping the review process, but by limiting the papers accepted to those that report only the very latest experimental or field research, are shorter in length, or meet other criteria. In this regard, producers of true *e-journals*, those journals that are not dependent on a paper equivalent, may succeed in improving the timeliness between the manuscript submission, the review process, and the availability or appearance of the article to readers.

Reports and other types of documents are often required of a researcher who has received funding from organizations or government agencies. The actual report format may be quite well defined by the organization so that it acquires all the information it needs for accounting purposes. Grant applications usually define reporting responsibilities.

Many scientists participate in prepublication databases and Web sites as a means of improving communication among researchers working in the same areas. A *preprint* is a manuscript that has not been published yet. The manuscript may be intended for presentation at a conference or for submission to a journal or some other publication. The PrePRINT Network http://www.osti.gov/preprint/ is an example of a Web site that links diverse preprint sites for the convenience of the searcher. Developed by the Office of Scientific and Technical Information of the U.S. Department of Energy, this network provides access to materials in the areas of physics, chemistry, mathematics, biology, environmental sciences, and other areas of science related to the Department of Energy's research interests. A collection of preprint sites, *arXiv.org e-Print Archive,* at http://arXiv.org/ provides links to research that has received support from the NSF.

Summary

Knowing how to conduct efficient research is a lifelong skill. Once you have received your undergraduate degree, you may use the information gleaned from this book to research educational, employment, research, and business opportunities.

Selecting a graduate school entails finding a school that offers a program of interest and finding a major professor. The *Peterson's* series and university Web pages are useful sources for finding specific information about university programs, entrance requirements, and costs. Electronic indexes, *Dissertation Abstracts,* and university Web pages are great places to start when trying to locate the names of scholars in your area of interest.

Once you find employment, you may be asked to pursue funding for your research. Federal organizations such as the NSF, NASA, and NIH provide useful starting points. In addition, the *Foundation Directory*, and the *Annual Register of Grant Support* can be used to identify potential funding sources.

Publishing is critical to the dissemination of scientific information. As you conduct your research, you will want to publish your results in books, journal articles, conference proceedings, and reports. Manuscript submission guidelines are normally included within an annual issue of a journal. To locate potential book publishers, survey the publishers of recent books in a related subject area or investigate the subject guides in *Literary Market Place* and *Books in Print.*

Most important, enjoy your research!

Note

1. *National Science Foundation: Overview of Grants and Awards.* http://www.nsf.gov/home/grants.htm (12 November 2001).

Appendix 1

Ten Tips for Efficient Library Research

1. Don't be afraid to ask a librarian for help! A reference librarian can make suggestions on the most helpful resources for your topic. He or she can also teach you how to search the resources most efficiently.

2. Allow plenty of time. Although the tips offered in this book should make your research more effective and efficient, it can still take time to locate and retrieve the complete text of the resources that you need. The perfect article may be available, but if it is not held at your library, you will need to have time to obtain it through interlibrary loan or document delivery.

3. Be sure you can state your topic in one or two sentences. If you are unclear on your research topic, you will probably end up having difficulty focusing your search and you will end up with lots of extraneous material. Select a topic that interests you—you will enjoy the research process more.

4. Spend a few minutes getting familiar with the topic and the resources available. Check general or subject-specialized encyclopedias and dictionaries to learn terminology and authors who write in that subject area. Check your library catalog to see if there are any books on the subject. Browse some peri-

odical indexes to determine what journals are covered, the dates of coverage, and the types of materials that are included.

5. You may need to refine your search. Refer to the controlled vocabulary of relevant resources to find suggestions on narrower, broader, or related terminology.

6. Select appropriate indexes, databases, and reference resources. If you are researching chemical reactions, consult resources that specialize in chemistry. On the other hand, don't be too restrictive. If your research topic crosses over more than one discipline—such as comparative psychology and animal behavior—make sure you check psychology resources in addition to zoological databases so that you can get both perspectives.

7. Use Boolean operators, proximity commands, and other advanced searching techniques to focus your on-line research. If you have difficulty setting up a good search strategy, check with the reference librarian.

8. Keep good notes and cite your resources. Remember to write down complete bibliographic citations for any research that you may want to cite in your paper. Provide your notes and bibliography in a standardized format such as the *American Psychological Association (APA)* or *Chicago.*

9. Evaluate, evaluate, evaluate! Think about the materials you are locating. Are the authors reputable? Have other authors cited the materials in their works? Is the material scholarly and based on sound research?

10. And the most important step: have fun!

Appendix 2

Sample Research Strategy

Now it's time to tie everything together. Suppose you have been asked to write a ten-page research paper for a class in ecology. How do you get started on your project? To illustrate the research strategies that we have proposed in this book, this appendix outlines a sample research strategy. The resources suggested here were retrieved using reference sources that are available in many academic libraries. As always, your own research strategy must be adapted to the resources that are available to you in your own library, in neighboring libraries, or through interlibrary loan. Ask your reference librarian for suggestions if you aren't sure what reference resources are available in your subject area.

Defining Your Topic

In order to research your topic accurately, you must have a good sense of *exactly* what you want to research. You should be able to state your research topic in one or two concise sentences.

Do not use a topic that is too broad, such as "ecology." You could not hope to discuss the entire field of ecology in one ten-page paper. Conversely, you can be too specific, too limiting or narrow, in your topic selection. In that case, your literature search may result in little or no information. It may be helpful to do some preliminary, general research to determine how much information is available on a particular topic. For more information on how to refine your search topic, refer to chapter 2.

For this example search strategy, we have decided to research the following topic:

Is it possible to predict the success or failure of a biological invasion by an introduced species?

Preliminary Research

It is normally helpful to examine some general reference sources first. Encyclopedias, dictionaries, and reference handbooks can provide help in selecting relevant terminology and may identify authors who are considered to be experts in the field of interest.

Specialized bibliographies may be useful during the early stages of research. Published lists of recommended resources are available in many subject areas. These resources normally provide *annotated* lists of specific encyclopedias, handbooks, directories, periodical indexes, and other reference materials that might be of use in a particular area. Check out the titles and the annotations. These short summaries are made by the author to indicate why the resource might be useful. Once you locate promising titles, you can go to the on-line catalog to see if they are available in your library. Your library staff may also provide selected subject bibliographies that describe appropriate resources that are available in your library.

Examples of published subject-specialized bibliographies include:

Hurt, C. D. *Information Sources in Science and Technology.*
3d ed. Englewood, CO: Libraries Unlimited, 1998.

Davis, Elisabeth B., and Diane Schmidt. *Using the Biological Literature: A Practical Guide.* 2d ed. New York: Marcel Dekker, 1995.

From the bibliographies listed here, you will find numerous ready-reference titles of interest, such as: *Encyclopedia of Environmental Studies, Encyclopedia of Environmental Biology, Encyclopedia of the Environment,* and *Dictionary of Ecology and Environmental Science.*

Perhaps most helpful are the suggested periodical indexes and abstracts. Here are listed specific periodical indexes for ecology: *Ecological Abstracts* and *Ecology Abstracts*. If those resources are not available, there are also suggestions for some general scientific periodical indexes, such as: *Biological Abstracts, Science Citation Index*, or *Zoological Record*.

While still in the process of locating background information, it might be helpful to examine some specialized encyclopedias. An article on "introduced species" in the *Encyclopedia of Environmental Biology* proves quite useful. This fourteen-page article provides a very nice overview of the topic. The author, Daniel Simberloff, has signed the article. It contains figures and photographs, a brief glossary, and a relatively current bibliography of additional resources. In addition to some basic information on the effects of introduced species, the article also provides some ideas for terminology and names of authors that might provide more information.

A quick check of Daniel Simberloff, the author of the encyclopedia article in biographical sources and in *Science Citation Index*, immediately shows that he is a prominent researcher. According to the 1998/99 edition of *American Men & Women of Science*, at the time of publication, he was a professor of biology at Florida State University and he specializes in ecology and mathematical biology.[1] Other authors also cite him extensively. All this indicates that Dr. Simberloff is one name to keep in mind as you proceed through your research.

As a starting point, it might help to check the library's on-line catalog for any of the resources that are listed in the bibliography of the encyclopedia article. These resources may provide useful information to get started on your research.

Remember to write down the citation for the encyclopedia article in the proper format *immediately* so that you will be able to return to the original source or cite it in your paper if necessary. There are many style formats available but, in this case, the professor has suggested that we use the *Chicago* style. A note card with the following *Chicago* citation is constructed as suggested in chapter 2:

> Simberloff, Daniel. "Introduced Species." In *Encyclopedia of Environmental Biology.* Vol. 2. San Diego: Academic Press, 1995.

While searching the reference stacks for the *Encyclopedia of Environmental Biology*, we also located a more recent five-volume encyclopedia entitled *Encyclopedia of Biodiversity*. A check of the index to this encyclopedia reveals a wealth of information relating to introduced species and biological invasions. The encyclopedia also contains another lengthy article by Daniel Simberloff titled "Introduced species, effects and distribution of." This thirteen-page article covers some of the same topics as his earlier article in the *Encyclopedia of Environmental Biology*, but additional material has been added. Not surprisingly, the bibliographies of both articles have a few citations in common; the *Encyclopedia of Biodiversity* article, however, contains references to materials that have been published since the earlier article was written. Both of these articles provide useful background information on the topic of biological invasions. Again, we make sure to document the source in the proper format:

> Simberloff, Daniel. "Introduced Species, Effects and Distribution of." In *Encyclopedia of Biodiversity*. Vol. 3. San Diego: Academic Press, 2001.

Before leaving the preliminary research sources, you may also want to check some ready-reference materials. In addition to the specialized bibliographies listed previously, there are several additional ways to locate ready-reference materials. Asking a reference librarian is always a good possibility. You can also go to the call number area for the broad topic (ecology) and scan the reference stacks. If you are interested in a particular type of reference work, such as a dictionary, you can look it up in the on-line catalog by performing a subject search with an appropriate subdivision, for example: ecology–dictionaries.

Specialized subject dictionaries, such as *A Dictionary of Ecology, Evolution and Systematics* by Roger Lincoln, Geoff Boxshall, and Paul Clark, may prove helpful along the way if you find terminology that you do not understand. Again, you will want to jot down the bibliographic information on this resource in case you need to refer to it again:

> Lincoln, Roger, Geoff Boxshall, and Paul Clark. *A Dictionary of Ecology, Evolution and Systematics*. New York: Cambridge University Press, 1998.

Identifying Key Concepts and Terminology

Now that you have a good background on the topic, it's time to start focusing the search. Before performing an actual subject search in the on-line catalog or the periodical indexes, identify appropriate subject headings. Begin by using the *Library of Congress Subject Headings*. Here, you will find some suggested standard subject headings:

> biological invasions
> plant invasions
> introduced mammals
> nonindigenous pests
> alien plants
> exotic plants

Be sure to check out suggested narrower terms or related terms. Just to be sure you have covered all your bases, it could also be useful to check the on-line catalog for several of the titles from the encyclopedia bibliography. If the books are listed, look to see what subject headings were used for those particular items. This may lead to some terms that you hadn't already considered.

Finally, you should check any periodical index thesauri or search guides that may be relevant. The 1997 *Zoological Record Search Guide*[2] lists several topics that might be helpful:

> dispersal by man
> introduction

The *BIOSIS Search Guide*[3] suggests:

> exotic (with a "see also" reference to the terms "foreign" or
>   imported")
> invasion (with a "see also" reference to "attack" or "infestation")

This should provide plenty of ideas for terminology with which to begin your search. You can adapt your search strategy if necessary as you proceed. For additional tips on selecting terms, refer back to chapter 4.

Locating Books

Using the Library of Congress subject headings listed in the previous section, go to the library's on-line catalog to see what books are available on this topic. A subject search of "biological invasions" produces approximately twenty records in our library catalog. Since this is a reasonable number of hits, it doesn't look like we will need to refine the search strategy yet. Several books of interest appear in the results list with one book in particular standing out:

> Drake, J.A., and others, eds. *Biological Invasions: A Global Perspective.* SCOPE Report no. 37. Chichester; New York: Scientific Committee on Problems of the Environment (SCOPE) of the International Council of Scientific Unions, 1989.

We immediately notice that this book was published in 1989, so it is definitely somewhat dated; the appropriateness of the title, however, is convincing enough to make it worth checking out. After pulling the book from the shelf, we note that it is actually a collection of contributed papers. Two chapters appear particularly relevant: chapter 11, "Ecosystem-level processes and the consequences of biological invasions" by P.S. Ramakrishnan and Peter V. Vitousek; and chapter 15, "Theories of predicting success and impact of introduced species" by Stuart L. Pimm.

Before actually taking the time to read the chapters in this book, see how the book rates using the evaluation system that was discussed in chapter 11. A quick check of *Science Citation Index* (via the *Web of Science* system) shows that the book, as a whole, has been cited extensively by other authors. Chapters 11 and 15 have also been cited. Looks good so far!

Now try checking out some book reviews. Using a general database, such as *Infotrac Expanded Academic Index,* provides citations to several different reviews of the book. After examining reviews from the journals *Evolution, BioScience, Science,* and *Ecology,* we begin to see some commonalities among the opinions of the reviewers. Generally, the reviews are quite favorable. While several reviewers admit that the book is somewhat lacking in hard answers, they acknowledge the major effort that went into this study.

The book certainly sounds like it will be an authoritative resource for research. A closer examination of the text shows that the chapter titled "Theories of predicting success and impact of introduced species" is very useful. This chapter provides an excellent overview of the theories specifically related to our topic. It also contains a fairly lengthy bibliography that might lead to additional helpful resources. A closer look at the bibliography also reveals a name that should be familiar, D. Simberloff, who was the author of the encyclopedia articles cited previously.

Remember, however, that the book was published in 1989 so you will want to locate more recent information on the topic, in case newer research has altered the theories that are presented here. One way to begin to check for more recent information is to go to *Science Citation Index* to look at the authors that cited this chapter in *their* research.

Before moving on to the next resource, be sure to cite the chapter in Chicago format for future reference. In this case, you should use the format for citing a chapter within an edited book:

> Pimm, Stuart L. "Theories of Predicting Success and Impact of Introduced Species." Chap. 15 in Drake, J.A., and others, eds. *Biological Invasions: A Global Perspective.* SCOPE Report no. 37. Chichester; New York: Scientific Committee on Problems of the Environment (SCOPE) of the International Council of Scientific Unions, 1989.

To locate books that are not available at your own library, there are several methods you can employ to find additional sources. *Books in Print* is a listing of books that are forthcoming, currently in print, or recently out of print. Another very useful locating tool that you will want to check is *WorldCat*, which is available through the *FirstSearch* system. *WorldCat* is a *union catalog* of more than 40,000 libraries from the United States and around the world.

WorldCat allows specific-subject searches using Library of Congress subject headings. An exact subject search of the subject heading "biological invasions" retrieves more than one hundred records. If any of these titles appear to be of interest, they can be requested from your interlibrary loan department. *WorldCat* probably will also include titles that are held at your own library, so be sure to check your on-line catalog before requesting them from interlibrary loan.

Locating Periodical Articles

Now we get to the "meat" of the research—the journal articles. These resources will provide specific information on the latest research. The first step for locating journal articles is always to select the most appropriate periodical indexes or abstracting services. As mentioned in chapter 8, there are a number of ways to locate appropriate indexes. Try asking the reference librarian or examining the specialized bibliographies that were mentioned previously in this sample search strategy. You can also check your library catalog or *Ulrich's International Periodicals Directory* for suggestions.

Using these guides, we find that there are several possible periodical indexes that might be appropriate for the topic. General science indexes like *Science Citation Index* could be helpful, or you may want to try out some of the biology indexes like *BIOSIS* or *Zoological Record*. In this case though, let's start with the index that sounds like it will be most specifically related to the topic: *Ecology Abstracts*. If there isn't enough material here, there is always the option to expand the search to some of the more general science databases.

First, look briefly at the information that describes *Ecology Abstracts*. The on-line version of this title is updated monthly and indexes more than 280 journals with coverage back to 1982. "Core" resources are indexed cover-to-cover with the other journals covered selectively; that is, the indexers would scan each journal and include citations to articles that are relevant to the field of ecology. A serials source list link from the Web site indicates which journal titles are covered selectively and which are comprehensively covered. The topics listed as being covered in this database look very appropriate.

Next, determine some relevant subject headings to use in this database. An on-line thesaurus is available and indicates that the subjects "introduced species" and "invasions" are authorized subjects for this database. Several related terms are also listed should it be necessary to expand the search.

This is a very specialized database, so it helps to be as specific as possible when constructing the search strategy. Since you want to concentrate on invasions of introduced species rather than natural invasions, first try a subject search of those two topics. Remember, whenever possible, use actual subject headings (or descriptors as they are called in this database) to perform the search. They will focus the search to retrieve records that the indexers felt discussed these topics as a major emphasis in the article. To search the descriptor (DE) field you would enter: DE=(invasions) and DE=(introduced species)

This search produces more than 175 hits. Many of the articles discuss individual species, but there are also a number of records that appear to be review-type articles. The search can be refined if necessary by modifying the search listed previously with a Boolean "and" and adding a keyword search of "predict*." This would limit the search to records that discussed predicting or prediction of the invasion of introduced species. It may still be useful to scan the records listed in the original search, however, since some good articles may not use the word "prediction" in the title or abstract of the paper.

The *Ecology Abstracts* search yields a number of helpful citations. Several jump out immediately. An article in *Science* magazine provides a recent overview of the status of biological invasion research and how successfully invasions can be predicted. Although the article is more of a news piece and it doesn't contain references, it does informally refer to some additional authors' works, including some by Daniel Simberloff. Of course, the reputation of *Science* as one of the core journals in scientific research gives a level of credibility to this overview article. Cite the article in Chicago format before moving on, in case you want to refer to it in your final paper:

Enserink, Martin. "Biological Invaders Sweep In." *Science* 285 (17 September 1999*)*: 1834-1836.

Ecology Abstracts also provides a large number of citations to scholarly research articles. A current article from *Trends in Ecology and Evolution* is one of the first scholarly articles to grab our attention. In this review of recent literature, the authors discuss whether there are factors that may be useful for determining whether or not a species will invade successfully. The article concludes with suggestions for additional research. The information contained here, particularly the relevant bibliography, will be a great starting point for our research. In addition, *Trends in Ecology and Evolution* is a refereed journal, which gives us greater confidence in its content. The full-text of this article was retrieved electronically so we must remember to include that in our citation:

Kolar, Cynthia S., and David M. Lodge. "Progress in Invasion
    Biology: Predicting Invaders." *Trends in Ecology and
    Evolution* 16 (2001): 199-204. Available from ScienceDirect.
    (6 December 2001).

While reading this article, another potentially important item is
found. At the beginning of the article, the authors refer to a 1958 book
by C.S. Elton titled *The Ecology of Invasions by Animals and Plants*.
What is interesting about this reference is that it was also cited in the
bibliography of both encyclopedia articles that were examined. The fact
that several authors have cited a book that was published many years
ago makes us pause. Checking *Science Citation Index* via *Web of Science*,
we discover that this particular book has been cited by others
more than 600 times! Obviously, this is a classic in this field that
should be examined:

Elton, Charles S. *The Ecology of Invasions by Animals and
    Plants*. London: Methuen, 1958.

Another article in *Trends in Ecology & Evolution* also addresses
the topic very specifically by discussing possible ways to predict invasions and analyzing whether current methods of prediction have been
successful. In addition, this particular article has been cited more than
175 times in *Science Citation Index*! The author, David M. Lodge, is
listed in *American Men & Women of Science* (20th ed.),[4] as a faculty
member specializing in invertebrate zoology and ecology at the University of Notre Dame. Lodge was also the co-author of the article discussed above. Once again, some now-familiar names are listed in the
bibliography of this article, including Simberloff and Drake. Although
this article was published in 1993, it looks like it is well worth
consideration:

Lodge, David M. "Biological Invasions: Lessons for Ecology."
    Trends in Ecology & Evolution 8 (1993): 133-137.

When checking *Science Citation Index* to see how often the Lodge
article has been cited, you may notice a reference to what appears to be
a brief comment on Lodge's article. This article adds some additional
insights:

Daehler, Curtis C., and Donald R. Strong Jr. "Prediction and
Biological Invasions." *Trends in Ecology & Evolution* 8
(1993): 380.

Let's look at one more example. In an article appearing in the *Canadian Journal of Fisheries and Aquatic Sciences*, the authors used some of the previously described methods and theories for predicting invasions and tried applying them to a specific case. The authors attempted to predict how many invertebrates from the Ponto-Caspian basin appeared to be likely to invade the Great Lakes-St. Lawrence River region through ballast-water exchanges. This is an interesting, practical approach to some of the reviews that we have located so far. The article is relatively recent (1998), is published in a refereed journal, has already been cited by several other authors, and contains references to authors that have already been noted:

Ricciardi, Anthony, and Joseph B. Rasmussen. "Predicting the
Identity and Impact of Future Biological Invaders: A Priority
for Aquatic Resource Management." *Canadian Journal of
Fisheries and Aquatic Sciences* 55 (1998): 1759-1765.

You will want to continue to read relevant periodical articles for your paper, but since you now know how to locate those with ease, let's move on to other types of resources. One final note on journal articles—it pays to start your research early. If your library does not subscribe to a particular journal title, you should be able to obtain a copy through interlibrary loan *if* you have allowed yourself sufficient extra time.

Locating Specialized Resources

Depending on the depth of your research and the amount of time available, you may want to check into research that has been published in the form of dissertations or in other specialized resources such as conference proceedings. It may be slightly more difficult to obtain a dissertation or a paper from a conference, but, if you are attempting to conduct exhaustive research on a topic, it is imperative that you consider this type of source.

Dissertation Abstracts is available from several vendors and provides an opportunity to try out some more advanced searching techniques. Your library may subscribe to Dissertation Abstracts through the FirstSearch system. FirstSearch is available in many libraries, so we will use its search protocols to illustrate the search strategy. Remember, though, if you have access to the on-line version of Dissertation Abstracts through another vendor, you can still perform a similar search. Just adapt the protocols and commands to that system.

In chapter 7, we described how searching Dissertation Abstracts might be somewhat frustrating because it tends to use very broad subject descriptors rather than a specific controlled vocabulary. Because of this, it provides a good example of how it may be necessary to adapt your search strategy as you work through the research process. Earlier in the research, we performed a subject search of "biological invasions" in the on-line catalog. Try the same search here. Using a keyword subject search, enter: biological invasion+.

This search produces about 100 records in Dissertation Abstracts. According to the help file in FirstSearch, you find that, if you don't give the computer any specific commands, it resorts to a "default" search. In FirstSearch, this means that the descriptor, the title, and the abstract fields were searched for the words "biological" and either "invasion" or "invasions." The + sign in FirstSearch tells the computer to look for the singular and the plural forms of the word. Since a Boolean operator wasn't used in the search strategy, the FirstSearch software assumed that you wanted to employ an "and" search between the two words. The search produces some relevant resources but also includes some false hits. All in all, this search wasn't too bad, but it can be improved.

Utilizing proximity searching techniques will force the software to locate the words in close proximity to one another. Again, using a keyword subject search, enter the phrase: biological n3 invasion+

This time, we are looking for all occurrences of the word "biological" within three words of the singular or plural form of the word "invasion," and the search produces twenty-four hits. These records look promising, but we also want to minimize the possibility that we missed some important citations.

What happens if the author of the dissertation didn't use the exact phrase "biological invasion?" You may be able to locate additional records if you searched for "exotic species" as well, but this brings up a rather sticky point. Not all biological invasions involve exotic species, and not all research on exotic species concerns the biological invasion aspect of their introduction. Therefore, it would seem logical to use the Boolean operator AND to limit the search to records that discussed both of these topics. If the literature on both topics is extensive, you will want to refine the search as much as possible. In the case of a multidisciplinary database like *Dissertation Abstracts,* however, an AND search may prove to be too restrictive. It may be that your results will be small enough that you will want to examine titles of any records that list *either* biological invasions or exotic species. In *Dissertation Abstracts*, try the following keyword subject search:

biological n3 invasion+
AND
exotic w1 species

The search statement is restrictive, resulting in very few hits. Let's try using an OR search with these two topics. Although this kind of search will undoubtedly produce some false hits, it is also more comprehensive, retrieving records that include either phrase.

The search of:

biological n3 invasion+
OR
exotic w1 species

produces more records than when searching for the phrase "biological invasion" alone. Since we have decided that we are most interested in the prediction of biological invasions, we can also try adding some variations of the term "prediction" to the previous search. Introducing terms such as "introduced species" or "non-indigenous species" are additional options.

If you decide to use one or more of the dissertations in your research, you must be able to cite it later. A dissertation by Gareth Russell would be cited as:

Russell, Gareth James. "Predicting the Appearance and
    Disappearance of Species: From Small Islands to Small
    Genera (Island Biogeography, Species Turnover, Extinction)."
    Ph.D. diss., University of Tennessee, 1996.

It is interesting to note that an examination of the complete citation for the dissertation by G. Russell reveals that his advisor was none other than Stuart L. Pimm, the author of the chapter in the first book that we located!

Conference and symposia proceedings are cited in many databases, including a number of periodical indexes. There are also methods for locating proceedings using specialized databases such as *PapersFirst* and *ProceedingsFirst*, which are available through the *FirstSearch* system; or *Conference Papers Index*, which is available from Cambridge Scientific Abstracts. Using these databases, you can locate papers from relevant conferences using the same kinds of search strategies that were used in the *Dissertation Abstracts* search. Remember, though, that papers from proceedings can be slightly more difficult to locate in the library. Refer back to chapter 8 for tips on how to locate these materials.

Locating Government Documents

As mentioned in chapter 10, the *Monthly Catalog* is considered the main index to government document materials and is always a good place to start when looking for government materials. The U.S. Government conducts and funds a great deal of research in the sciences and publishes a lot of information in this area. The *Monthly Catalog* is available from several vendors and is sometimes listed under a variation of the name so; if you are unable to locate it in your library, check with a reference librarian.

If you do not have access to the exact subject headings used in the *Monthly Catalog* database, begin the search with a quick and dirty search strategy. The *Monthly Catalog* is a multidisciplinary database and doesn't include abstracts, so there is not a lot of searchable text. With that in mind, begin the search in a very general way. Start with a keyword search of: "biological invasions."

Several records are retrieved. Glancing through the citations, there doesn't appear to be any documents that look particularly useful for this research paper, but it is interesting to note that the word "nonindigenous" appears in many of the citations. Consider revising the search to see what records are retrieved with a keyword search of "nonindigenous or nonindigenous." This search produces more citations and several are references to hearings related to the *National Invasive Species Act of 1996*. The lengthy House and Senate Hearing Reports may contain useful details that can be used as background for the paper.

While the citations are on the screen, it is highly recommended that you write down the SuDoc classification number as well as the citation information. These will help locate the material later on:

> U.S. Congress. House. Committee on Transportation and Infrastructure; Subcommittee on Water Resources and Environment. *H.R. 3217, the National Invasive Species Act of 1996: Hearing before the Subcommittee on Water Resources and Environment and Subcommittee on Coast Guard and Maritime Transportation, Committee on Transportation and Infrastructure, House of Representatives,* 104th Cong., 2d sess., 17 July 1996.

SuDoc number: Y4.T68/2:104-68

> U.S. Congress. Senate. Committee on Environment and Public Works. Subcommittee on Drinking Water, Fisheries, and Wildlife. *National Invasive Species Act of 1996: Hearing before the Subcommittee on Drinking Water, Fisheries, and Wildlife of the Committee on Environment and Public Works, United States Senate, One Hundred Fourth Congress, second session, on S. 1660 to Provide for Ballast Water Management to Prevent the Introduction and Spread of Nonindigenous Species into the Waters of the United States, and for Other Purposes, September 19, 1996.* 104th Cong., 2d sess., 19 September 1996.

SuDoc number: Y4.P96/10:S.HRG.104-746

Locating Statistical Information

Statistical information would provide backup information on the severity of some biological invasions. There are a number of places where statistical information can be located. Statistical information is often poorly indexed in regular periodical indexes or abstracts, but several databases are specifically designed to help locate statistical information.

FactSearch, available through the *FirstSearch* system, indexes statistical information in journals, documents, newspapers, newsletters,

Congressional information, and Internet sites. *FactSearch* is multidisciplinary in its coverage. Keep the search statement relatively general—you probably won't need to narrow down the search with words
like "predicting" or "prediction." A search of "biological w1 inva-
sion+" in *FactSearch* produces approximately ten records. In addition
to the actual statistics, the records retrieved also provide the exact citation where the data originally appeared. A search of "introduced n3
species or exotic w1 species" also provides some statistics that might
be used as background information.

Another resource for statistical information is *Statistical Universe*.
This database includes the print resources *Statistical Abstract of the
United States*, *American Statistics Index*, *Statistical Reference Index*,
and *Index to International Statistics*. A keyword search on "exotic species" retrieves a citation for a recent article in the *Chronicle of Higher
Education*. The full-text of the article is not included as part of *Statistical Universe*, so you will need to locate a copy on the periodical
shelves. The article contains numerous figures within the text of the
article as well as a table that details the "Annual costs of species introduced into the United States":

> McDonald, Kim A. "Biological Invaders Threaten U. S.
> Ecology." *Chronicle of Higher Education* 45 (12 February
> 1999): A15-A16.

The World Wide Web

As we mentioned in chapter 9, most printed publications have to pass at
least a minimal level of editing and selection before they are published,
whereas Web sites can be published by anyone who has access to
server space. A lot of really good information is available through open
access Web sites, but this is definitely an area that requires the use of
good evaluation skills.

Before jumping into a general, broad-based Web search, try using
some of the specialized Internet search tools that may be available. You
may want to start your Internet search by checking with your local library to see if it has compiled a relevant subject bibliography that contains Internet sites. The *Cambridge Scientific Abstracts* databases, such
as *Ecology Abstracts*, also index some relevant Web sites. These sites

are checked monthly to ensure that they are still active, and they are se-lected by the editors of *Cambridge Scientific* as sites that should be helpful for their users.

NetFirst, a database that is available through the *FirstSearch* sys-tem, is another possible resource for good sites. *NetFirst* indexes Inter-net sites that have been evaluated and selected according to a definitive collection policy. You can feel comfortable that the sites you locate through *NetFirst* have passed an initial level of evaluation. The sites are also checked regularly to be sure they are still active. Searches of "in-troduced species" and of "exotic species" produce several records on the general topic, but they don't appear to be relevant for our project. A search of "invasive species," however, does lead to a potentially useful Web site called *invasivespecies.gov*.

We immediately notice the ".gov" domain in the URL. This tells us that this is a government Web site. Click on the link and then apply some additional evaluative criteria to the site. Although *NetFirst* takes us to a specific section of this Web site, let's begin by going to the "In-vasive Species Home" link. The nicely organized Web page promi-nently displays a definition of an invasive species and provides a "con-tact us" link. You quickly find a link to the agency that has authored the page: NBII. If you don't know anything about NBII, you can go to the home page, which is readily accessible from this site. There, you will find that NBII stands for the National Biological Information Infra-structure. A link to "about NBII" reveals that it is a collaborative initia-tive of government, academic, and private industry partners working to provide electronic access to biological information. A detailed listing of the organizations involved in this program is provided on the Web site. Sounds good so far! Now let's check out the timeliness of the page. The main page was last updated the day before we accessed it.

The site certainly looks reputable, but does it contain relevant in-formation? Several items look interesting. The site is nicely organized with links to individual species, specific geographic areas, organiza-tions, photographs, and many other resources. The law and regulations link provides information on relevant pending legislation. Another link provides resources relating to the economic impact of invasive species, journal articles, and reports that list specific costs. The *invasivespe-cies.gov* site is a wonderful resource for very specific or more general-ized background information on the topic of biological invasion. If we decide to cite any of the specific documents that are linked to this Web site, we will need to cite them individually. In the meantime, be sure to cite the main page in its proper format:

National Biological Information Infrastructure.
invasivespecies.gov. http://www.invasivespecies.gov/
(6 December 2001).

The use of specialized databases has produced an interesting Web site, but there should be more information out there. It is now time to perform an "open" Web search to see what is available. In order to keep the search as relevant as possible, employ some advanced searching techniques. For instance, the *Hotbot* search engine allows an "exact phrase" search or a search on "all the words." We can also limit the search to Web sites that are part of the ".edu" or ".gov" domains. This should help to ensure that we retrieve sites that will be of sufficient quality for academic research.

Since it is necessary to be as specific as possible when performing an open Internet search, let's begin with a search of "predicting biological invasions" (all the words) and limit it to the .edu or .gov domains.

The search results include links to several college course syllabi, but it also retrieves some promising sites. In the results list, we note an article on biological invasions that appears in the summer 1997 issue of *Issues in Science and Technology*. Don Schmitz and Daniel Simberloff wrote the article. We've already checked on the qualifications of Daniel Simberloff, so we know that he is a qualified researcher. At the end of the article, we also find information about the authors. At the time of publication, Don Schmitz worked as a wetland and alien plant coordinator for the Florida Department of Environmental Protection.[5]

This Web site links back to the home page for *Issues in Science and Technology*, where we learn that this journal is published by the National Academy of Sciences, the National Academy of Engineering, and the University of Texas at Dallas. The "about us" link on the home page indicates that this journal is a "forum for discussion of public policy related to science, engineering, and medicine."[6] The home page has a link to the current issue and to previous issues going back to 1996. A quick check of *Ulrich's* shows that there is a printed version of this journal, and its publication dates back to 1984. The journal is refereed and is covered by a vast number of periodical indexes. All this information leads us to believe that this is a credible journal.

Schmitz, Don C., and Daniel Simberloff. "Biological Invasions: A Growing Threat." *Issues in Science and Technology.* Summer 1997. http://www.nap.edu/issues/13.4/schmit.htm. (6 December 2001).

Bibliography

In the preceding pages, we have attempted to provide specific examples on how to conduct library research. Obviously, you will need to adapt the steps to the resources available to you in your own library; the basic steps, however, remain very similar.

The final step (other than writing your paper of course!) is to compile your footnotes and bibliography. You must always be sure to give proper credit to an author's original work. If you refer to his or her research in any way, you must cite him or her in your bibliography. You should select the style recommended by your professor or your publisher. If you are not given a particular style to follow, you may select any of the scientific style formats—just be sure you pick one and stick with it to ensure proper consistency.

Figures A2.1-A2.2 provide examples on how to cite different kinds of resources. These bibliography page samples illustrate the Chicago format for many of the resources that were mentioned in this exercise.

Notes

1. "Simberloff, Daniel S." in *American Men and Women of Science 1998-99,* 20th ed. vol. 6 (New York: R.R. Bowker, 1998), 924.

2. *Zoological Record Search Guide.*™ (Philadelphia: BIOSIS, 1997), C88.

3. *BIOSIS Search Guide* (Philadelphia: BIOSIS, 1997), C-101, C141.

4. "Lodge, David Michael" in *American Men and Women of Science, 1998-99,* 20th ed. vol. 4 (New York: R.R. Bowker, 1998), 1027.

5. *Issues in Science and Technology, About IST.* http://www.nap.edu/issues/about.html (12 December 2001).

6. Don C. Schmitz and Daniel Simberloff. "Biological Invasions: A Growing Threat." *Issues in Science and Technology.* Summer 1997, http://www.nap.edu/issues/13.4/schmit.htm (17 Nov. 2001).

References

Daehler, Curtis C., and Donald R. Strong Jr. "Prediction and
    Biological Invasions." *Trends in Ecology & Evolution* 8
    (1993): 380.
Davis, Elisabeth B., and Diane Schmidt. *Using the Biological
    Literature: A Practical Guide*. 2d ed. New York:
    Marcel Dekker, 1995.
Drake, J.A., and others, eds. *Biological Invasions: A Global
    Perspective*. SCOPE Report no. 37. Chichester; New York:
    Scientific Committee on Problems of the Environment (SCOPE)
    of the International Council of Scientific Unions, 1989.
Elton, Charles S. *The Ecology of Invasions by Animals and Plants*.
    London: Methuen, 1958.
Enserink, Martin. "Biological Invaders Sweep In." *Science* 285
    (17 September 1999): 1834-1836.
Hurt, C. D. *Information Sources in Science and Technology*. 3d ed.
    Englewood, CO: Libraries Unlimited, 1998.
Kolar, Cynthia S., and David M. Lodge. "Progress in Invasion Biology:
    Predicting Invaders." *Trends in Ecology and Evolution* 16 (2001):
    199-204. Available from ScienceDirect.(6 December 2001).
Lincoln, Roger, Geoff Boxshall, and Paul Clark. *A Dictionary of
    Ecology, Evolution and Systematics*. New York: Cambridge
    University Press, 1998.
Lodge, David M. "Biological Invasions: Lessons for Ecology."
    Trends in Ecology & Evolution 8 (1993): 133-137.
McDonald, Kim A. "Biological Invaders Threaten U. S. Ecology."
    Chronicle of Higher Education 45 (12 February 1999): A15-A16.
National Biological Information Infrastructure. *invasivespecies.gov*.
    http://www.invasivespecies.gov/ (6 December 2001).
Pimm, Stuart L. "Theories of Predicting Success and Impact of
    Introduced Species." Chap. 15 in Drake, J.A., and others, eds.
    Biological Invasions: A Global Perspective. SCOPE Report no. 37.
    Chichester; New York: Scientific Committee on Problems of the
    Environment (SCOPE) of the International Council of Scientific
    Unions, 1989.
Ricciardi, Anthony, and Joseph B. Rasmussen. "Predicting the Identity
    and Impact of Future Biological Invaders: A Priority for Aquatic
    Resource Management." *Canadian Journal of Fisheries and
    Aquatic Sciences* 55 (1998): 1759-1765.

Figure A2.1. Sample bibliography page in *Chicago* format.

Russell, Gareth James. "Predicting the Appearance and Disappearance of Species: From Small Islands to Small Genera (Island Biogeography, Species Turnover, Extinction)." Ph.D. diss., University of Tennessee, 1996.

Schmitz, Don C., and Daniel Simberloff. "Biological Invasions: A Growing Threat." *Issues in Science and Technology.* Summer 1997. http://www.nap.edu/issues/13.4/schmit.htm. (6 December 2001).

Simberloff, Daniel. "Introduced Species." In *Encyclopedia of Environmental Biology.* Vol. 2. San Diego: Academic Press, 1995.

———. "Introduced Species, Effects and Distribution of." In *Encyclopedia of Biodiversity.* Vol. 3. San Diego: Academic Press, 2001.

U.S. Congress. House. Committee on Transportation and Infrastructure; Subcommittee on Water Resources and Environment. *H.R. 3217, the National Invasive Species Act of 1996: Hearing before the Subcommittee on Water Resources and Environment and Subcommittee on Coast Guard and Maritime Transportation, Committee on Transportation and Infrastructure, House of Representatives,* 104th Cong., 2d sess., 17 July 1996.

U.S. Congress. Senate. Committee on Environment and Public Works. Subcommittee on Drinking Water, Fisheries, and Wildlife. *National Invasive Species Act of 1996 : Hearing before the Subcommittee on Drinking Water, Fisheries, and Wildlife of the Committee on Environment and Public Works, United States Senate, One Hundred Fourth Congress, second session, on S. 1660 to Provide for Ballast Water Management to Prevent the Introduction and Spread of Nonindigenous Species into the Waters of the United States, and for Other Purposes, September 19, 1996.* 104th Cong., 2d sess., 19 September 1996.

Figure A2.2. Sample bibliography page in *Chicago* format.

Glossary

Because some library jargon is inevitable in a book of this sort and some words may be used with other than their general usage, we provide this glossary of some terms as they are used in library research.

Abridged dictionary A dictionary that does not include all the words in a language (as opposed to *unabridged*).

Abstract A summary of the content of a bibliographic item.

Abstracting journal An index that provides abstracts to the items indexed, in addition to the citations. Also called an *abstracting service*.

Abstracting service An index that provides abstracts to the items indexed, in addition to the citations. Also called an *abstracting journal*.

Academic library A library that serves a college or university.

Access point Any searchable piece of information in a bibliographic record.

Accession number The sequential number given to a particular biblio-graphic item in a database, either print or electronic. In a print abstract-ing service, it is the link between the index(es) and the *main entry*.

Aggregator A *vendor* that provides access to the electronic publica-tions, such as journals or newspapers, of several different publishers.

Almanac A ready-reference source, generally annual, that includes both current and historical statistical information and general facts.

AND The *Boolean operator* that specifies that *both* of two concepts must be present in a record for it to satisfy the conditions of the search statement; AND narrows or focuses a search strategy.

Annotated With additional comments, as an annotated bibliography, i.e., one containing evaluative material in addition to the citation.

Annotation An additional comment. Personal annotations in e-books can reappear if the same user "checks out" the same book again. Please do not make annotations in *p-books*!

Annual A serial issued on a yearly basis, such as an *almanac* or a *yearbook*.

Atlas A collection of flat maps bound in book form.

Bibliographic information The information necessary to identify a particular library item, generally the information included in a *biblio-graphic record* or a *citation*.

Bibliographic record The basic unit of library catalogs. The represen-tation of a library item in a catalog or database with fields for the bib-liographic and subject information describing the item.

Bibliography 1. A collection of citations at the end of a paper. 2. A published collection of references to materials on a particular topic.

Blind review The strictest form of peer review, in which neither the author(s) nor the reviewer(s) know the identity of the other, to avoid bias.

Book number The second part of a *call number*, which follows the *class stem* and identifies a particular item within a broader classification.

Boolean commands Commands used to instruct a computer database to combine concepts in the desired manner. The three Boolean commands are AND, OR, and NOT. Also called *Boolean operators.*

Boolean operators Commands used to instruct a computer database to combine concepts in the desired manner. The three Boolean commands are AND, OR, and NOT. Also called *Boolean commands.*

Broader Term In a *thesaurus*, a more generic term that includes the narrower terms beneath it. Often abbreviated "*BT.*"

Browser A software program that displays text, graphics, and multimedia from the Internet.

BT Broader term.

Call number A unique number assigned to a library item to indicate its location within the library. A call number is based on one of several classification systems and consists of a *class stem* and a *book number*.

Card catalog A manual file of cards containing bibliographic records representing materials available in the library and indicating the location of the item. Virtually all research libraries have replaced the card catalog with an *on-line public access catalog,* at least for recent materials.

Case sensitivity The response of a search engine to commands typed in uppercase versus lowercase.

Case specificity The requirement that a search result must match the case (uppercase, lowercase, or initial caps) of the search statement.

Catalog A listing of bibliographic records representing materials available in the library, either manual (a card catalog) or electronic (an on-line public access catalog), searchable by various *access points* and indicating the location of the item in the library.

Citation A reference to the source of information included in a work, if not the author's own, generally either in a *footnote* or an *endnote*.

Citation index A particular form of index in which the references in an article's bibliography are indexed and searchable.

Cited reference search A search for more recent references that cite, in their bibliographies, a previously identified item.

Claim A statement in a patent that specifies the unique aspects of the thing being patented.

Class stem The first part of a *call number*, which indicates the item's position in the classification scheme.

Classification system A system of organizing objects or information based on similar characteristics. In library use, a system for arranging library materials, usually by subject, and generating a *call number* to locate the item on the shelf. Classification systems discussed in this book are the Dewey Decimal system, the Library of Congress system, the National Library of Medicine system, and the Superintendent of Documents (SuDocs) system.

Conference proceeding A published record of the papers presented at a conference. Proceedings may include only the abstracts of the papers or they may contain the full-text and graphics of the paper.

Controlled vocabulary A listing of authorized terms for assigning subjects in an index or database. Controlled vocabularies ensure that a consistent subject term is used when several possibilities exist for the same concept.

Copyright A form of intellectual property protection extended to written and creative works.

Corporate author An organization such as a society or governmental body that publishes an item under the organization's name, without attributing it to a personal author.

Crawler A program that scans the World Wide Web for new sites, which are then added to the database of a *search engine*. Also called a *spider*.

Curriculum vita An extended resume listing a scientist's or engineer's education, publications, honors, etc. Often called simply a *vita.*

Database A collection of information organized into records and fields for ease of retrieval. In library research, the term normally refers to an electronic index to some portion of the literature.

Depository library A library that receives documents free of charge for access by the public. *GPO* depository libraries may be selective or regional. The Patent Depository Library program is separate from the GPO depository library program.

Descriptor An indexing term or *subject heading* selected from a controlled vocabulary.

Dewey Decimal Classification System A classification system developed by Melvil Dewey and used by most public and school libraries. Call numbers in the Dewey Decimal System can be recognized by a stem consisting entirely of numbers, specifically, three numbers (000 to 999) followed by a decimal point.

Directory 1. A ready-reference source listing persons or organizations, with addresses or telephone numbers. 2. A hierarchically arranged index to Web sites.

Dissertation A formal account of original research prepared as part of the requirements for a doctoral degree.

Document delivery A system in which documents are purchased from a commercial source on demand for a fee.

Domain 1. The last element in a *URL*, which identifies the type of site or country of origin where the Web site was produced. Examples are .edu for educational institutions or .it for Italy. 2. A collection of Web sites on a single topic from a single publisher.

Double-blind review A form of peer review in which neither the author(s) nor the reviewer(s) know the other's identity to avoid bias.

Drop-down menu A device in many search screens that allows the user to select from a set of options. The user generally clicks on an arrow, which displays the list of options, then scrolls down to the desired choice. Clicking on the choice, then selects it for use. Also called a *pull-down menu.*

e-book A book in which the full content is delivered across the Internet.

e-journal A journal in which the full content is delivered across the Internet. E-journals may be electronic versions of print journals or may exist only electronically.

Endnote A note at the end of a work giving additional information or the source of material.

Fair use Guidelines developed to determine how much of another author's work can be used without obtaining formal permission; proper credit must be given for *any* use of another author's words or ideas (see *plagiarism*).

False drop A record retrieved by an electronic database that satisfies the criterion of the search statement as written, but that does not satisfy the intent of the search. Also called a *false hit.*

False hit A record retrieved by an electronic database that satisfies the criterion of the search statement as written, but that does not satisfy the intent of the search. Also called a *false drop.*

FAQ Frequently asked questions. A Web page that lists a variety of common questions about the subject of the Web site.

Federal document A government document issued by the federal (United States) government.

Field A component of a *record* that contains a specified type of information, as the "author" field or the "publication year" field.

Field searching Limiting a search to a particular *field* in a database to focus results or eliminate *false hits*.

Footnote A note at the bottom of a page giving additional information or the source of material.

Free-text searching Searching for a term anywhere within a *record*, rather than within a specified *field*.

Full-text resource An electronic resource that provides the full-text of the item indexed, in addition to the indexing information. Depending on the resource, "full-text" may mean simple text without illustrations or page images complete with graphics.

Gazetteer A listing of place names and locations.

Government document A document issued by a government agency. Generally refers to those documents produced by the *Government Printing Office* and distributed to *depository libraries* for access by the public. Can also refer to similar documents published and distributed by state governments.

Government Printing Office The agency of the federal government that publishes federal government documents and makes them available to the public through depository libraries or for sale.

GPO *Government Printing Office.*

Grant In science and engineering, an award from a government or charitable organization to support research.

Gray literature Materials that have been published but are not indexed in standard indexing tools and are therefore difficult to identify and locate. Some government agency reports fall into this category.

Handbook A ready-reference source containing compilations of useful data in a specific subject area.

Hit A *record* retrieved from an electronic database using a search statement.

Hot links Links embedded in text that allow a user to move between documents on a Web site. Also called *hyper-links*.

http Hyper-text transfer protocol.

Hyper-links Links embedded in text that allow a user to move between documents on a Web site. Also called *hot links*.

Hyper-text transfer protocol (http) The Internet protocol that allows the transfer of hyper-linked text and multimedia on the World Wide Web.

Icon A simple graphic image that stands for a hyper-link on a Web page.

ILL Interlibrary loan.

Intellectual property protection Legal safeguards to protect the rights of creators to profit from their own intellectual endeavors. The forms of intellectual property protection in the United States are *copyright, patents, trademarks*, and *trade secrets*.

Interlibrary loan A service provided by most libraries to borrow materials needed by their patrons from other libraries.

International Standard Book Number A unique number that identifies a specific edition of a specific book by a specific publisher. Used primarily in the book trade, the *ISBN* is not the same as a call number.

International Standard Serial Number A unique number that identifies a specific periodical or other serial publication. Different formats (print vs. electronic) of the same serial title may or may not have different *ISSNs*.

Internet A network that allows computers connected to it to exchange information using a set of standard protocols.

Internet Service Provider A company or organization that provides a connection between a computer and the Internet.

IP address A number that uniquely specifies a particular computer to the Internet.

ISBN *International Standard Book Number.*

ISP Internet service provider.

ISSN *International Standard Serial Number.*

Journal A periodical aimed at a scholarly audience.

Juried Of a journal, one in which the articles are reviewed by experts in the field prior to publication to certify their value to the field of study, also called *peer-reviewed* or *refereed.*

Keywords Words entered into an electronic database to identify relevant items.

Laws Rules enacted by Congress (federal) or legislatures (state), as opposed to *regulations.*

LC The Library of Congress, or, the classification system used by the Library of Congress and most large academic libraries.

LCSH The Library of Congress Subject Headings.

Legislation The collective term for a body of *laws.*

Library of Congress Classification The classification system used by the Library of Congress and most large academic and research libraries. The Library of Congress system is based on subjects and can be recognized by a stem consisting of one or two (rarely three) letters followed by a number from 1 to 9999.

Library of Congress Subject Headings The *controlled vocabulary* used by the Library of Congress to assign subject headings to books. Libraries that use the Library of Congress classification system to organize their collections generally also use the Library of Congress Subject Headings in their catalogs.

Limit To restrict a search result based on one or more criteria such as date or language.

Limiter A *field* used to limit a search result.

Location In library terms, the physical building or area where an item is shelved. Since the stem of a call number can be in any of several places within the library (e.g., the book, periodical, or reference stacks), the location should also be noted when copying a call number from the catalog.

Lower case Non-capital letters. In the days of movable type, small letters were in the lower drawers of the type case. The adjective is low-ercase.

Magazine A periodical aimed at a general audience.

Main entry An entry containing all information about a bibliographic item, which is referred to by a variety of index entries. In an abstracting journal, the entry that contains all of the bibliographic information and the abstract that is accessed through a variety of indexes. In the days of card catalogs, the main entry was usually on the author card, with abbreviated information on the title and subject cards.

Manual A ready-reference source containing procedural information.

Map A graphic representation of a portion of the earth or other celestial object.

Mark With reference to electronic databases, to specify one or more items from a list to be saved for downloading or printing.

Microcard A microform in which the images are stored on a piece of thin cardboard of varying size. Also called *microopaque*.

Microfiche A microform in which the images are stored on a rectangular piece of transparent plastic approximately 4 by 6 inches. Microfiche with particularly high reduction ratios is called *ultrafiche*.

Microfilm A microform in which the images are stored sequentially on a roll of plastic film.

Microform A generic term for media that store information as reduced photographic images.

Microopaque A microform in which the images are stored on a piece of thin cardboard of varying size. Also called *microcard*.

Narrower term In a *thesaurus*, a more specific term within the broader term above it. Often abbreviated *NT*.

National Library of Medicine Classification A classification system used by most medical libraries.

Natural language searching Searching an electronic database using normal English phrasing, including *stop words*. Natural language searching is not supported by all databases.

NOT The *Boolean operator* that eliminates any item containing the specified term from final consideration. The NOT command narrows or focuses the search results, but must be used with caution.

NT Narrower term.

OCLC Online Computer Library Center, Inc. A worldwide library organization of over 40,000 libraries. OCLC provides a variety of services to libraries, including shared cataloguing and preservation. It is the organization responsible for *WorldCat*.

On-line Public Access Catalog A computerized, searchable listing of bibliographic records representing materials available in the library and indicating the location of the item.

OPAC On-line Public Access Catalog.

Open Internet All the World Wide Web that is available free to anyone, as opposed to sites that are transmitted over the Internet, but available only to subscribers.

OR The *Boolean operator* that specifies that either of two or more terms (or both) may appear to satisfy the search statement. The OR command is used for related terms and is used to broaden a search.

Patent A form of intellectual property protection extended to inventors for their inventions.

Patent and Trademark Office The agency of the United States government responsible for issuing patents and trademarks.

P-books Books that are published in paper format.

Peer-reviewed Of a journal, one in which the articles are reviewed by experts in the field prior to publication to certify their value to the field of study. Also called *juried* or *refereed*.

Periodical A publication issued at regular intervals (weekly, monthly, quarterly) under a single title, generally with multiple articles in each issue; a generic term for magazines and journals.

Phrase searching Searching for two or more words as a phrase, rather than separate words.

Plagiarism The act of claiming the work of another as one's own.

Polyglot dictionary A dictionary providing equivalent words in multiple languages.

Preprint A draft of a manuscript distributed over the Internet for comment prior to submission to a journal.

Primary author The first author among those listed for an article or book.

Projection The method of translating features on a curved surface (the earth) onto a flat map. All projection systems introduce distortion.

Protocol A set of standards that allows information to be transmitted and received over the Internet.

Proximity command Database-specific commands used to define the number of words between or order of two search terms in a search statement. Also called *proximity operators*.

Proximity operators Database-specific operators used to define the number of words between or order of two search terms in a *search statement*. Also called *proximity commands*.

Proximity searching A method to focus a search or reduce *false hits* by specifying that two search terms must be within a specified distance of or order relative to each other through the use of *proximity operators*.

Proxy server A system that authenticates users from an *Internet Service Provider* so that they can access the subscriptions of another institution.

PTO The United States Patent and Trademark Office.

Pull-down menu A software feature that provides a list of available options for a function, whereby a user clicks on a term or an arrow and the list is displayed. The user clicks on the desired choice to select. Also called a *drop-down menu*.

Ready-reference An information need that can be answered with a simple fact or illustration, which can generally be answered by consulting a single reference work, or such a reference work.

Record A representation of an item in a database. Records are composed of *fields*, each of which holds similar information in different records.

Refereed Of a journal, one in which the articles are reviewed by experts in the field prior to publication to certify their value to the field of study, also called *juried* or *peer-reviewed*.

Reference A citation included in a footnote or endnote.

Reference librarian A professional trained to guide users to the information they need. Your best resource in an academic library.

Regional depository library A *depository library* that receives all documents available through the *GPO* depository library program. There may be no more than two regional depositories per state.

Regulations Rules passed by an agency of the Executive Branch of government to carry out the intent of *laws.*

Related term In a *thesaurus*, another authorized term that has any relationship to the indexed term other than broader or narrower. Often abbreviated *RT.*

Relevancy In the Internet world, a computerized process that ranks a set of results for presumed usefulness based on various measures of the prominence of the search terms in the result.

RT Related term.

Scope note A note in a *thesaurus* indicating how a particular term is used in the context of the database.

Search engine A collection of software that scans the World Wide Web, indexes the sites found, and allows a user to identify sites based on desired keywords.

Search statement The final form of the request to an electronic database to search for relevant items. Also called a *search strategy.*

Search strategy The final form of the request to an electronic database to search for relevant items. Also called a *search statement.*

Search terms Words entered into an electronic database to identify relevant items.

Selective depository library A *depository library* that receives only a portion of the government documents available.

Serial A generic term for an item issued under the same title over time without a foreseeable end. Serials include *periodicals, annuals,* and irregulars (those issued at unpredictable intervals).

Server A computer attached to the Internet that holds data that can be viewed by other computers attached to the Internet.

Spider A program that scans the World Wide Web for new sites, which are then added to the database of a *search engine.* Also called a *crawler.*

Stacks Collectively, the shelves where books or journals are housed.

Stop words Words so common that they are not indexed by computer databases. Common stop words are "the," "is," and "of."

Subdirectory A set of computer files grouped within a larger computer file.

Subdivision A restriction of a subject heading narrowing the format, time period, or other aspect of the broader heading. Also called a *sub-heading.*

Subheading A subdivision of a *subject heading* narrowing the format, time period, or other aspect of the broader heading. Also called a *sub-division.*

Subject heading An access point in a catalog selected from a *controlled vocabulary* indicating the major subject of a work.

SuDocs number The number in the *Superintendent of Documents (SuDocs) classification* system assigned to an individual government document. Unlike most other classification systems, which are based on subjects, the SuDocs system is based primarily on the issuing agency. A SuDocs call number can be recognized by the presence of both letters and numbers, and by the presence of a colon in the number.

Superintendent of Documents classification A system of classifying government documents based primarily on the issuing agency, generally abbreviated as *SuDocs.*

Thesaurus In indexing, the listing, print or electronic, of *controlled vocabulary*, including unused synonyms, broader or narrower terms, and related terms.

Thesis A formal account of original research prepared as part of the requirements for a master's degree.

Trade magazine A periodical issued for practitioners of a field of study, generally containing primarily news items and advertisements.

Trade secret A form of intellectual property protection in which the details of a process are not divulged or registered.

Trademark A form of intellectual property protection that identifies and distinguishes the goods or services of one party from those of another.

Truncation The process of searching for variant endings of a search term by using a designated symbol to represent additional letters at the end of the stem.

UF Use for.

Ultrafiche A microfiche in which the images are reduced at a higher ratio than normal microfiche.

Unabridged dictionary A dictionary that attempts to include all of the words in a language, both current and historical.

Unary system A system of *Boolean commands* employed by some Internet *search engines* where a plus sign (+) indicates that a term *must* appear in the results and a minus sign (-) indicates that the term *may not* appear.

Union catalog A "super-catalog" containing the contents of several other catalogs, usually from several institutions, and showing any and all locations where the item is located.

Universal resource locator A phrase that specifies the unique location of a resource on the Internet. *URLs* consist of an Internet protocol such as "http://" followed by a host name and a series of subdirectories.

Upper case Capital letters. In the days of movable type, capital letters were in the upper drawers of the type case. The adjective is uppercase.

URL Universal Resource Locator.

USE In a *thesaurus*, the indication that the given term is not the authorized form for a concept, with a cross reference to the authorized form.

Use for In a *thesaurus*, an indication that the given term is the authorized form for a concept. Often abbreviated *UF*.

Vanity press A publisher who will publish anything for a fee.

Vendor In the library world, a company from whom the library purchases resources. The same electronic resource may frequently be purchased from several different vendors, each of which provides slightly different features.

Venn Diagram A graphic representation of Boolean algebra showing the way in which logical sets interact.

Vita A *curriculum vita.*

Web site A collection of related documents published on the *World Wide Web.*

Webliography A bibliography of *Web sites.*

Wildcard A symbol used in electronic searching to represent varying letters within a search term, such as "wom*n" for "woman" or "women."

World Wide Web An Internet protocol that allows the display of text, graphics, and multimedia.

WWW The *World Wide Web.*

Yearbook An annual ready-reference source that describes recent activity in a field or provides updates to an existing source.

Bibliography

Listed below are some research tools that are available in most academic libraries or on the open Internet. Individual science-specific periodical indexes, encyclopedias, and other reference sources are too numerous to mention here and are covered nicely in other publications. Included here are general, multidisciplinary sources mentioned in the book that may be useful in scientific and technical research.

American Men and Women of Science, 20th ed., (1998/1999), New York: R.R. Bowker, 1998.

First published in 1906 as *American Men of Science,* this directory provides brief biographical information on living scientists and engineers. It is updated about every three to four years. To get information on deceased persons, you have to go back to earlier volumes.

American Statistics Index. Washington, D.C.: Congressional Information Service, 1973-present.

This resource provides a comprehensive index to statistical publications of the United States government, both items distributed through the depository process and those that are not. The index is published monthly in two parts, indexes and abstracts, and the monthly issues are cumulated in a bound volume annually. Statistical publications are indexed by subject, names, titles, and report numbers. Libraries may also subscribe to microfiche versions of the full documents indexed, though the cost may be prohibitive for smaller libraries; check with your reference librarian.

Annual Register of Grant Support: A Directory of Funding Sources, 35th (2002) ed. Medford, N.J.: Information Today, 2001.

> This annual directory provides information on sources of grant funding. The individual entries provide contact information, eligibility requirements, and application details with dates. The book is arranged by broad subject areas and has subject, organization, and geographical indexes.

Biography and Genealogy Master Index (BGMI). Detroit: Gale Research, 1975-present.

> The main index to biographical reference resources. BGMI allows fast and easy access to the resource name, volume, and year where an individual's biography will appear.

Book Review Digest, New York: H.W. Wilson, 1905-present

> An index to book reviews that have appeared in popular and scholarly periodicals. Although not a full-text resource, many of the citations include excerpts from the review.

Book Review Index, Detroit: Gale Research, 1965-present.

> Indexes book reviews appearing in popular, trade, and scholarly journals. Organized by author and indexed by title.

Books in Print, New York: R.R. Bowker, 1948-present.

> This annual compilation provides basic bibliographic citations and purchasing information for North American books that are currently in-print, forthcoming or recently out-of-print. Many libraries also provide access to *Books in Print* electronically; check with your librarian.

Cambridge Scientific Abstracts, http://www.csa.com/

> *Cambridge Scientific Abstracts* is a group of databases that are searched using the same interface. A number of databases discussed in the text and in this appendix, such as *Aquatic Sciences and Fisheries Abstracts, Conference Papers Index, Ecology Abstracts, GeoRef,* and *Oceanic Abstracts,* are accessible through *Cambridge Scientific Abstracts.* There are a number of databases devoted to the sciences, engineering, and medicine. Your library may subscribe to some or all of the Cambridge Scientific databases. [Note: The address given above is the generic *Cambridge Scientific Abstracts* Web site; your library may have IP-based links further down in the Web site.]

Chicago Manual of Style, 14th ed., Chicago: University of Chicago Press, 1993.

> Regularly used in the field of publishing, this manual provides instruction on spelling, punctuation, and formatting of notes and bibliographies. Particularly helpful for guidance on preparing a manuscript for publication.

Code of Federal Regulations. National Archives and Records Administration, Office of the Federal Register. Washington, D.C.: GPO, 1938-present.

The *Code of Federal Regulations* (CFR) is the compilation of all active federal regulations. It is divided into 50 "titles," which are further divided into chapters, parts, and sections, with the sections being the individual regulations. The CFR is reissued annually, with the issuance of the various titles staggered, thus, in July, some titles will be dated the current year and others the previous year. Between publications of successive titles, updates are published in the *Federal Register.* The CFR is also available on the Internet through GPO Access.

Conference Papers Index, Cambridge Scientific Abstracts, 1982-present. http://www.csa.com

Indexes the papers and poster sessions that have been presented at major scientific meetings worldwide. [Note: The address given above is the generic *Cambridge Scientific Abstracts* Web site; your library may have IP-based links further down in the Web site.]

Congressional Record. Washington, D.C.: GPO, 1873-present.

The *Congressional Record* and its predecessors have published the "Proceedings and Debates of the Congress" every day that one or both Houses is in session since 1789. Included in the *Congressional Record* are the full texts and status of bills, addresses by members of Congress, and anything else that a Congressperson wishes to include. The *Congressional Record* is *not* a verbatim transcript of the activities of Congress—Congresspersons may insert remarks that were not made on the floor and may edit the content of the *Congressional Record* after publication. An annual cumulation is prepared for the permanent record, and details of the daily and annual *Congressional Record* may vary; document the version that you cite. The *Congressional Record* is also available on the Internet through GPO Access.

Contemporary Authors, Detroit: Gale Group, 1962-present.

This series provides basic biographical and literary information on current writers of note. Each volume contains a cumulative index to all volumes in the set.

CRC Handbook of Chemistry and Physics: A Ready-Reference Book of Chemical and Physical Data, 82nd ed. Ed.-in-Chief David R. Lide. Boca Raton, Fla.: CRC Press, 2001.

The subtitle accurately describes this compendium, which has been published annually since 1913. In addition to information on chemicals, it also includes extensive tables of conversion factors, astronomical data, and mathematical formulae.

Current Biography Yearbook. New York: H.W. Wilson, 1940-present.

Published annually, this resource contains moderately lengthy biographical essays on persons of fame throughout the world. Articles include a photo and a brief bibliography.

Dissertation Abstracts International. Ann Arbor, Mich.: UMI, 1952-present.

Bibliographic information and the abstract of doctoral dissertations published in the United States are indexed in *Dissertation Abstracts International*. In 1966, the publication split into an A part and a B part, with science and engineering dissertations in part B. *Dissertation Abstracts International* is published by UMI, which maintains microfilm copies of all dissertations abstracted, and copies can be ordered from them. *Dissertation Abstracts* is also available online through various subscription services (check with your reference librarian). The online version includes dissertations back to 1861 and many master's theses, as well.

Expanded Academic ASAP™ Gale Group, 1980-present.
http://web6.infotrac.galegroup.com

This multidisciplinary index covers scholarly journals in science and technology, social sciences, and humanities. Coverage is back to 1980 and many article citations link to full-text.

FactSearch. OCLC. 1984-present. http://newfirstsearch.oclc.org/
One of the *FirstSearch* databases, this resource provides full-text excerpts with statistical information on a wide variety of topics. Each excerpt contains the full citation for the original source. [Note: The address given above is the generic FirstSearch login screen; your library may have IP-based links further down in the Web site.]

Federal Register. National Archives and Records Administration, Office of the Federal Register. Washington, D.C.: GPO, 1936-present.

The *Federal Register* is the source of information on pending regulations. It is indexed quarterly and cumulated annually into the *Code of Federal Regulations*. It is published daily, Monday through Friday, except federal holidays, and arranged by agency. Within each agency are listed rules, proposed rules, and notices, if there is activity on that day. In 2000, there were over 83,000 pages published in the *Federal Register*. The *Federal Register* is also available on the Internet through GPO Access.

Foundation Directory. New York: The Foundation Center, 1960-present.

This annual directory provides basic information on nongovernmental and nonprofit organizations that give large amounts in grant funding. Indexes include subject, geographical area, and foundation name.

GPO Access. U.S. Government Printing Office, Superintendent of Documents. http://www.access.gpo.gov/su_docs/index.html

> *GPO Access* is the master gateway to U.S. government information. A wide variety of legislative, executive, and judicial documents and information is available through *GPO Access.*

GrayLit Network: A Science Portal of Technical Reports. Department of Energy, Office of Scientific and Technical Information. http://www.osti.gov/graylit/

> The GrayLit Network provides access to non-classified technical reports of the U.S. Departments of Defense and Energy, the EPA, and NASA that have not been published through normal publication channels and may not be readily available in libraries. It is searchable by keyword.

Index to the U.S. Patent Classification. Department of Commerce, Patent and Trademark Office. Washington, D.C.: GPO, 1977-present.

> An annual index to the terms within the hierarchical *Manual of Classification* of patents. The *Index* is also available on the Internet at the PTO Web site.

Internet Public Library. University of Michigan, School of Information. Sponsored by Bell & Howell Information and Learning. http://www.ipl.org/

> The *Internet Public Library* is a directory of the Internet organized along library lines. There is a reference area that leads to a variety of Web sites with useful information. The Online Texts Collection [http://www.ipl.org/reading/books/] links to over 18,000 full-text documents. These documents are either classic texts that are no longer under copyright or government documents. They can be searched by author, title, or Dewey Decimal classification. (Remember that 500 is science and 600 is technology.)

JSTOR®. http://www.jstor.org/

> JSTOR is a digital archive of scholarly journals in a number of fields. All issues of the journals are digitized cover-to-cover from the beginning to a "moving wall" approximately five years before the current year. Journals in JSTOR may be browsed by issue or searched in full-text.

Lexis-Nexis® Academic Universe. Reed-Elsevier, 2001. http://web.lexis-nexis.com/universe

> *Lexis-Nexis Academic Universe* is a collection of databases including many newspapers from around the world; business, legal, and medical information; and basic reference sources. *Academic Universe* is the best source of current newspaper indexing.

Lexis-Nexis® Statistical Universe. Reed-Elsevier, 2001.   http://web.
lexis-nexis.com/statuniv/.

> *Statistical Universe* is an electronic database of statistical informa-
> tion, including parts of the *American Statistical Index, Statistical Refer-
> ence Index,* and *Index to International Statistics.* It can be searched by
> keywords. Many of the resources include the complete tables online, and
> searches can be limited to those with full-text.

Library of Congress Subject Headings, 23rd ed. Washington, D.C.:
Library of Congress Cataloging Distribution Service, 2000.

> This five-volume work lists the controlled vocabulary used by the Li-
> brary of Congress in its catalog, with broader, narrower, and related terms.
> It is thus a "master thesaurus" for all fields of study. Formerly issued
> irregularly, it is now issued annually.

Manual of Classification. Department of Commerce, Patent and
Trademark Office. Washington, D.C.: GPO.

> The *Manual of Classification* lists the hierarchical system used to
> classify patents. It is issued in loose-leaf form and updated as necessary.
> The *Manual of Classification,* with definitions, is also available on the
> Internet at the PTO Web site.

Monthly Catalog of United States Government Publications. U.S. Government Printing Office, Superintendent of Documents. Washington, D.C.: GPO, 1895-present.

> The *Monthly Catalog of United States Government Publications,* or
> "MoCat," is a listing of all documents published by the U.S. Government
> Printing Office and distributed to depository libraries. Within MoCat,
> documents are listed by their SuDocs classification number, which is
> based on the issuing agency. It is indexed annually, and commercial
> cumulative title and subject indexes are available. Since 1996, entries in
> MoCat have been abbreviated, giving primarily bibliographic information,
> as cataloging information is available through a variety of sources. The
> *Monthly Catalog* is available on the Internet through GPO Access and a
> number of other sources.

National Science Foundation: Where Discoveries Begin.
http://www.nsf.gov

> The National Science Foundation is one of the largest funders of sci-
> entific research. The Foundation's Web site includes information on grant
> programs, as well as news releases on discoveries funded by the NSF.

*National Technical Information Service: The Central Source for Scien-
tific, Technical, and Business-Related Government Information.*
Department of Commerce, National Technical Information Service. http://www.ntis.gov

The National Technical Information Service (NTIS) publishes many U.S. government scientific and technical reports that are not distributed through the Government Printing Office depository program. Over 400,000 documents are now available through NTIS and may be searched at their Web site. These documents may be available in your library, through Interlibrary Loan, or by purchase from NTIS.

NetFirst. OCLC. http://newfirstsearch.oclc.org/
One of the *FirstSearch* databases, this resource indexes current Internet resources. The database is updated daily and locations are checked regularly for accuracy. [Note: The address given here is the generic FirstSearch login screen; your library may have IP-based links further down in the Web site.]

Notices of Funding Availability. Department of Agriculture, Office of Community Development. http://ocd.usda.gov/nofa.htm
Notices of Funding Availability (NOFA) is a database of grant announcements published in the *Federal Register*. The database may be searched by broad keywords ("environment," "research and development") and by agency.

OCLC FirstSearch®. http://newfirstsearch.oclc.org/
FirstSearch is a family of databases in many fields that are searched using the same interface. A number of databases discussed in the text and in this appendix, such as *Books in Print, Dissertation Abstracts, Fact-Search, the Monthly Catalog, PapersFirst,* and *WorldCat,* are accessible through *FirstSearch.* There are also a number of databases specifically related to the sciences, engineering, and medicine including *Agricola, General Science Abstracts, Geobase, Inspec,* and *Medline.* Your library may subscribe to some or all of the FirstSearch databases. [Note: The address given above is the generic *FirstSearch* login screen; your library may have IP-based links further down in the Web site.]

Official Gazette of the United States Patent and Trademark Office: Patents. Department of Commerce, Patent and Trademark Office. Washington, D.C.: GPO, 1872-present.
This weekly publication lists all patents issued during the previous week, giving bibliographic information, one drawing, and one claim for each patent. Patents are arranged numerically. An annual index to the *Official Gazette* is prepared; patents issued since 1976 are indexed electronically through CASSIS, a database provided by the Patent and Trademark Office, and all patents are now available on the Internet at the PTO Web site.

The On-Line Books Page. Ockerbloom, John Mark, ed., 1993-2001. http://digital.library.upenn.edu/books/
This site provides a gateway to over 15,000 full-text books on the Internet. It is searchable by author, title, and Library of Congress classification.

Oxford English Dictionary, 2nd ed. Prepared by J.A. Simpson and E.S.C. Weiner. Oxford: Clarendon Press, 1989.

    The *Oxford English Dictionary*, or OED, is the most comprehensive of all dictionaries of the English language. In addition to definitions, the OED traces the history of words through their use in various literary and other contexts. In addition to the print volumes, the OED is also available electronically; check with your reference librarian for availability in your library.

PapersFirst. OCLC. 1993-present. http://newfirstsearch.oclc.org/

    One of the *FirstSearch* databases, this resource indexes papers that have been presented at scientific meetings worldwide. [Note: The address given above is the generic FirstSearch login screen; your library may have IP-based links further down in the Web site.]

Peterson's Graduate Programs in Engineering and Applied Sciences; Peterson's Graduate Programs in the Biological Sciences; Peterson's Graduate Programs in the Physical Sciences, Mathematics, Agricultural Sciences, the Environment and Natural Resources, Princeton, NJ: Peterson's. 35th ed., 2001.

    This series contains basic information on graduate programs, faculty, and the universities where various programs are offered. Indexed by institution and organized by major, it also contains useful tips on applying to and funding graduate school.

PrePRINT Network. U.S. Department of Energy. Office of Scientific and Technical Information. http://www.osti.gov/preprint/

    This government resource provides access to scientific and technical preprints; manuscripts that may have been accepted for publication but are not yet published or manuscripts that are being circulated for comment prior to publication. In most cases the full-text of the preprints is available without a subscription. An alerting service is also available.

ProceedingsFirst. OCLC, 1993-present. http://newfirstsearch.oclc.org/

    One of the *FirstSearch* databases, this resource indexes the British Library Collection of papers presented at symposia, conferences, congresses, expositions, workshops, and professional meetings. [Note: The address given above is the generic FirstSearch login screen; your library may have IP-based links further down in the Web site.]

Project Gutenberg. 1971-2001. http://promo.net/pg/

    Project Gutenberg is the largest and oldest online text database. Works in Project Gutenberg are classic works in all fields.

Publication Manual of the American Psychological Association, 5th ed., Washington, D.C.: American Psychological Association, 2001.

    A style regularly used in scientific publications, the APA manual provides information on spelling, punctuation, capitalization, and the preparation of manuscripts.

Science Citation Index. Philadelphia: Institute for Scientific Information, 1961-present.

This large, interdisciplinary index to scientific research is particularly useful for its wide coverage of scientific journals. Cited reference searching allows you to locate related articles that were published after the resource in question. Retrospective volumes were published covering the literature back to 1955. The *Science Citation Index* is available electronically through *Web of Science.*

Social Sciences Citation Index. Philadelphia: Institute for Scientific Information, 1969-present.

A major index to literature in the social sciences. Cited reference searching allows you to locate related articles that were published after the resource in question. The *Social Sciences Citation Index* is available electronically through *Web of Science.*

Statistical Abstract of the United States: The National Data Book. Census Bureau. Washington, D.C.: GPO, 1878-present.

The *Statistical Abstract of the United States,* or "Stat Abs," is a compendium of information on the United States on a wide variety of topics. Most of the tables are derived from federal government documents, but some are also abstracted from private sources. Some tables break the information down to the level of individual states. Each table has a footnote indicating the source of the information; this source can generally be consulted for additional information on the subject. Note that references in the index are to table number, not page. The *Statistical Abstract of the United States* is also available on the Internet through GPO Access.

Statistical Reference Index. Bethesda, Md.: Congressional Information Service, 1980-present.

The *Statistical Reference Index* is a companion set to *American Statistics Index* and provides indexing of private and state statistical information. It is issued in the same format as ASI and has many of the same indexes. The original documents are published on microfiche, though some items under copyright protection cannot be reproduced; approximately 90% of the documents were provided on microfiche in 2000.

Thomas: Legislative Information on the Internet. Library of Congress. http://thomas.loc.gov.

Thomas provides up-to-date information on activities of the U.S. Congress. In addition to providing the full-text and status of bills currently under consideration by Congress, *Thomas* provides links to other Congressional resources.

Ulrich's Periodicals Directory, 39th ed., New Providence, N.J.: R.R. Bowker, 2001.

This is the place to begin when you want to find out about a particular periodical. The directory is not an index to articles in journals but, rather, gives information about the journal itself; publication data and subscription address and price. Perhaps the greatest value to researchers is in the descriptions of where a particular journal has been indexed and its listing of refereed journals.

United States Code, 1994 ed. Washington, D.C.: GPO, 1995.

The *United States Code* is a compilation of all laws currently in effect, arranged by subject. The *U.S. Code* is divided into 50 titles, which do not correlate with the 50 titles of the *Code of Federal Regulations.* The *U.S. Code* has been published at intervals since 1926. Currently, the entire *U.S. Code* is reissued every six years. Annual supplements keep the *U.S. Code* updated between editions. The *U.S. Code* is also available on the Internet through GPO Access. Some private companies publish annotated versions of the *United States Code* that include cross-references to court decisions and other useful information; your library may or may not subscribe to one of these private versions.

United States Government Manual. National Archives and Records Administration, Office of the Federal Register. Washington, D.C.: GPO, 1935-present.

The *Government Manual* provides information on the agencies of all three branches of the federal government. Each entry provides the address, phone number, and Web address of the agency; names of major officials; and the history and purpose of the agency. Since it is issued annually, names of officials may have changed, so it is generally a good idea to confirm those with a more current source. The *Government Manual* is also available on the Internet through GPO Access.

United States Patent and Trademark Office: An Agency of the United States Department of Commerce. Department of Commerce, Patent and Trademark Office. http://www.uspto.gov/

The official Web site of the United States Patent and Trademark Office provides information on searching and filing applications for patents and trademarks. Patents back to number 1 in 1790 are provided (a free plug-in viewer is required to view images), as are all forms and manuals necessary to search the databases.

United States Statutes at Large. Washington, D.C.: GPO, 1873-present.

The *Statutes at Large* is an annual bound set of all laws passed by Congress during a session. It is arranged chronologically, so laws on similar subjects are not necessarily together. For laws in effect at a given time, arranged by subject, see the *United States Code.* Both the *Statutes* and the *Code* are cross-referenced to each other. Laws passed prior to 1873 were published by private printers. Larger libraries may have reproductions of these compilations.

Web of Science. Institute for Scientific Information.
http://webofscience.com/

This large, multidisciplinary database provides Web access to *Science Citation Index, Social Sciences Citation Index,* and *Arts and Humanities Citation Index.* Cited reference searching allows you to locate related articles that were published after the resource in question.

Who's Who in... New Providence, N.J.: Marquis Who's Who.

This series is well known for its basic biographical entries on persons of note. Entries are relatively brief and provide information on the person's life, career, and honorary awards. There are a number of titles within the series that specialize in a geographic area (*Who's Who in America, Who's Who in the Midwest,* etc.) or in a field of study (*Who's Who in Science and Engineering,* etc.).

WorldCat. OCLC. http://newfirstsearch.oclc.org/

WorldCat is the combined catalogs of over 40,000 libraries in 76 countries with over 41 million records in over 400 languages. *WorldCat* can be used to identify materials pertinent to your research that are not owned by your library or to see if a nearby library has the materials you need. [Note: The address given here is the generic FirstSearch login screen; your library may have IP-based links further down in the Web site.]

Index

Page numbers for illustrations and figures are in italics.

abstracting services, 78, 84-86, *87-88*, 89, *90;* electronic, 89-91. *See also* periodical indexes

abstracts, 43, 78, 83; in electronic databases, 89; in periodical abstracting services, 86, 89

academic libraries, 1-3, 73, 108

access points, 57-58, 89

almanacs. 23. *See also* ready reference sources

AND command, 38-39, *40, 41, 42,* 46, 169; implied, 51; in World Wide Web searches, 104-105; troubleshooting search results, 45, 47. *See also* Boolean commands

articles, locating, 27, 77-78, 83-92, 164-67

atlases, 24

bibliographic records, 59-60, 65-66, 76

bibliographies, specialized,158

bibliography, 8-11, 14, 166, 175; in encyclopedia articles, 21-22, 159, 161; management

software, 67-68; sample in *Chicago* format, 176-77

biographical resources, 130-31, 145, 159

Biography and Genealogy Master Index, 130

book databases, 61-68, 162-63; electronic, 70-72; publishers catalogs, 69. *See also Books in Print; WorldCat*

Book Review Digest, 136

Book Review Index, 136

book reviews, 136, 139, 145, 162

books, 57, 162-63; databases, 68-69; electronic, 70-72; evaluation, 136, 139, 145, 162; locating, 61-68

Books in Print, 68-69, 76, 151, 163

Boolean commands, 38-43, 46, 61, 156, 165, 168-69; in advanced searches, 91, 96, in searching government information, 119; in World Wide Web searches, 103-106

Boolean operators. *See* Boolean commands